PLANETS
OURS ∧ND OTHERS

From Earth to Exoplanets

PLANETS
OURS AND OTHERS
From Earth to Exoplanets

Thérèse Encrenaz

Paris Observatory, France

edp science

orld Scientific

Published by

World Scientific Publishing Co. Pte. Ltd.

5 Toh Tuck Link, Singapore 596224

USA office: 27 Warren Street, Suite 401-402, Hackensack, NJ 07601

UK office: 57 Shelton Street, Covent Garden, London WC2H 9HE

Library of Congress Cataloging-in-Publication Data
Encrenaz, Thérèse, 1946–   author.
    Planets : ours and others : from Earth to exoplanets / Thérèse Encrenaz (Paris Observatory,
France).
    pages cm
    Includes index.
    ISBN 978-9814525152 (pbk. : alk. paper)
    1. Planets. 2. Extrasolar planets. I. Title.
    QB601.E527 2013
    523.4--dc23

                                                2013021878

British Library Cataloguing-in-Publication Data
A catalogue record for this book is available from the British Library.

Originally published in French as "**Les Planètes**" by EDP Sciences.
Copyright © EDP Sciences 2011. A co-publication with EDP Sciences, 17, rue du Hoggar, Parc
d'activités de Courtaboeuf BP 112, 91944 Les Ulis Cedex A, France.

This edition is distributed worldwide by World Scientific Publishing Co. Pte. Ltd., except France.

Typeset by Stallion Press
Email: enquiries@stallionpress.com

Printed in Singapore by Mainland Press Pte Ltd.

Thérèse Encrenaz, born in 1946, is a Senior Scientist at the Centre National de la Recherche Scientifique. She works at LESIA (Laboratory of Space and Instrumental Studies for Astrophysics) at Paris Observatory. Her expertise is the study of planetary atmospheres, in particular by remote sensing analyses, using space and ground-based data. She has been involved in many space missions (Vega, Phobos, Galileo, ISO, Mars Express, Venus Express, Rosetta). She is the author of over 250 articles in refereed journals and a dozen popular books. She received the silver medal of CNRS in 1998, the Janssen medal of the Astronomical French Society in 2007, and the David Bates medal of the European Geophysical Union in 2010.

Acknowledgements

I wish to thank Fabienne Casoli and Athena Coustenis who carefully read the manuscript. I also thank Marc Ollivier and Athena for their help in preparing the figures. Finally, I wish to thank all the colleagues who have allowed me to include some of their work in this book.

Contents

Foreword

This book by Thérèse Encrenaz is among the first of a new series "Introduction to...", aimed at addressing a scientific question in simple and accessible language, far from the jargon of specialists. This question — the nature, origin and evolution of planets — is especially timely, as we now know over 900 planets orbiting nearby stars, in addition to our eight solar system planets.

Planetary astronomy is a science almost as old as civilization, since the ancient Babylonians and Assyrians already knew the motions of planets. Following Newton, celestial mechanics, devoted to the study and the prediction of celestial motions, has developed as far as becoming "the queen of sciences"; recent developments of this discipline, previously thought to be outdated, are surprising and spectacular. In contrast, due to the lack of appropriate observational means, progress in the physical study of planets has been quite slow after the plethora of discoveries accumulated in the XVIIth century by Galileo, Huygens and Cassini. Only half a century ago, we still knew almost nothing about the nature of the planets and their atmospheres, not to mention their satellites. Then, thanks to big ground-based telescopes and radio-telescopes, Earth-orbiting observatories and space probes, our knowledge has exploded. A new discipline, comparative planetology, has emerged from these often unexpected discoveries acquired over the last five decades. It opens for us exciting horizons about the origin and evolution of planets, including our own Earth. In addition, the discovery of extrasolar planets — one of the greatest discoveries of today's astronomy — opens a new dimension in the study of planets, and promising perspectives about the quest for extraterrestrial life. Our present understanding of planets and exoplanets is far from being definitive: the great variety of solar system planets and satellites, and, even more so, the multiplicity of exotic planetary systems discovered so far raise many unanswered questions.

It could seem impossible to synthesize in a few pages such a rich and complex topic. Still, the author has succeeded in this task, thanks to her pedagogic skills and her deep knowledge of the subject; in fact, she is one of the scientists who have developed planetology in France. This book will interest not only the educated layman, but also specialists of the field. The reader will easily understand the text, dense but clear, and helped by beautiful images coming from the space probes and the big telescopes. This well written book, on a topic of great importance, should encounter a durable success.

James Lequeux
Emeritus Astronomer, Paris Observatory

Introduction

What is a planet? The question may look silly, as its answer seems obvious. Still, the definition of a planet has evolved significantly over the centuries. The Greeks gave the name of "planets", i.e. "wandering objects", to the celestial bodies moving in the sky with respect to the stars. In Antiquity, only visible planets were known, and we still use their latin translation: Mercury, Venus, Mars, Jupiter and Saturn. In the XVIth century, following the Copernican revolution, planets were defined as bodies in orbit around the Sun, and the Earth was added to the list, followed two centuries later by Uranus. At the beginning of the XIXth century, the discovery of the biggest asteroids made the situation more confused. Astronomers soon understood that a new class of objects had been discovered, the main belt asteroids located between Mars and Jupiter. As more and more of these objects were expected to be discovered, they were withdrawn from the official list of planets. After Neptune's discovery in 1846, the list included thus eight objects.

A new surprise came in 1930: Pluto, a distant object orbiting the Sun, was discovered beyond Neptune's orbit. Logically, it became the ninth planet. It kept this status until 2006, when the International Astronomical Union (IAU) decided to remove it from the list. What has happened in between? Following hunting campaigns with larger and larger telescopes, a new class of objects was discovered: the transneptunian objects (TNOs). Like Pluto, they are located beyond Neptune's orbit, in a region of the solar system called the Kuiper Belt. Their existence had been suspected a few decades ago, on theoretical bases, by two astronomers, K. Edgeworth and G. Kuiper; it allowed us to explain, in particular, the origin of comets with low inclinations and short periods. As discoveries of TNOs accumulated, it has become obvious that Pluto is only one of the largest members of this family; its large size made its discovery possible well before the other TNOs. A proof of Pluto's origin is given by the fact that many TNOs have the same revolution period as Pluto, which is exactly 1.5 times Neptune's period: Pluto and Neptune are said to be in 3:2 resonance, and the same applies to all other TNOs (called Plutinos) with the same period.

In 2003 came the end of Pluto's fate as a planet: Eris, a TNO as big as Pluto, or possibly even bigger, was discovered. It is much farther from the Sun that Pluto, which explains why it was not discovered earlier. Now appears the evidence: the Kuiper Belt contains thousands of objects of this kind, possibly bigger than Pluto, which remain to be discovered. Thus, it becomes impossible to keep Pluto as a planet, unless the risk of having to enlarge the list of planets indefinitely is taken. This is why, very logically, the list of planets has been reduced to the eight planets known before 1930. The most massive TNOs, as well as the biggest asteroid, Ceres, have received the label of "dwarf planets" (see Insert in Annex).

Our eight planets can be divided into two distinct classes, very different in nature. In the vicinity of the Sun, within 2 AU (the AU is the astronomical unit, i.e. the mean Sun–Earth distance or more precisely, the semi-major axis of the Earth's orbit), the four terrestrial planets are characterized by a relatively small radius and a high density: they are also called the rocky planets. Beyond 5 AU, the four giant planets are found, very large but with small densities, all surrounded by a ring system and a large number of satellites. We will see later how this fundamental difference between the two classes of planets can be explained in the light of the formation scenario of the solar system.

Let us go back to the definition of a planet. It looked simple at the beginning; after the discovery of extrasolar planets — or planets orbiting other stars — it is no more the case. The discovery, since 1995, of hundreds of exoplanets surrounding nearby stars has been a true revolution for astronomy. The solar system is no longer a unique phenomenon, even if the planetary systems found so far look very different from the one we know. Thus, the concept of a planetary system, and hence the concept of a planet, needs to be revisited. The definition given by the IAU in 2006 is not very clear for the layman (see Insert). The object, in orbit around a star, must have cleared the space around its orbit (this is to exclude asteroids or Kuiper Belt type objects). This definition will possibly evolve in the future, as new exotic objects will be discovered. Let us try to define what are, in our view, the essential characteristics of a planet, those which make it specific.

At the center of the planetary system, we find the star, which has a thermonuclear energy source. As the star evolves, it transforms its hydrogen (entirely produced during the Big Bang, in the so-called primordial nucleosynthesis) into helium; then, C, N and O are formed, then heavier and heavier elements including metals and, in particular, iron. These nuclear reactions are at the origin of the radiation emitted by the Sun and the stars. In contrast, planets do not have this energy source, because their mass and internal temperature are not sufficient for the thermonuclear cycle to start. Using theoretical modeling, the critical mass needed to start this cycle can be estimated: it is about 13 times the mass of Jupiter. If the mass is above 80 times the Jovian mass, the object falls in the star class. Between 13 and 80 Jovian masses, it belongs in an intermediate class called "brown dwarfs". Its mass is sufficient to start the first thermonuclear cycle which transforms hydrogen into deuterium; the temperature is a few million degrees. But this is not sufficient to start the next step, which is the formation of helium: a temperature of ten million degrees would be needed. Brown dwarfs are "aborted" stars, so to speak, which were interrupted early in their thermonuclear cycle.

With no thermonuclear source, planets still have some internal sources of energy, but those have nothing in common in terms of intensity. For giant planets, the internal energy is a leftover of the gravitational energy accumulated during the accretion phase of the planets; for telluric planets, the radioactive elements present in the interior feed an internal source which is responsible for volcanism and plate tectonics. However, the sources only add to the main energy source, which comes from the solar (or stellar) radiation.

Beyond the definitions of experts, the first characteristic of a planet appears clear: the light emitted by the planet does not come from its own interior, but comes from the light of the Sun, or its host star. This light can be either reflected or scattered, at the same wavelength as the solar radiation, or absorbed by the planet and converted into thermal heat; in this case the radiation peaks at longer wavelengths. Similar to every object in the Universe, planets have an intrinsic radiation (called the blackbody radiation) associated with their temperatures. In the case of the solar-system planets, this temperature is at most a few hundred K and peaks in the infrared range. An equilibrium takes place between the absorbed solar radiation and the thermal emission corresponding to the equilibrium temperature of the planet. The contribution of the internal energy may also have to be included (of gravitational origin for giant planets, of radioactive origin for terrestrial planets). The closer the planet to its host star, the more efficient is the heating; its equilibrium temperature decreases as the planet's distance to the star increases. Let us now take the case of an exoplanet. For a given star, there is a distance at which the equilibrium temperature will be above 0°C in a range allowing water, if present, to be in liquid form. This case is of most interest for us: could such planets look like the Earth and host life? This fundamental question motivates our interest in planets and exoplanets.

The purpose of this book is to try to characterize planets, both in their general entity and in their diversity. Starting from our planet Earth, then describing the diversity of solar system planets, we will try to show how a few essential parameters (the distance to their star, their mass, density, obliquity, rotation period...) determine their physico-chemical properties (chemical composition, thermal and cloud structure, atmospheric circulation, seasonal effects, climate...). We will then be in a better position to explore the new field opened to us, the exoplanets. Starting from the experience acquired from solar system planets, we will try to imagine what their composition and structure may be, on the basis of the few parameters we know. We keep the quest for extraterrestrial life still in mind: could it exist or have existed, in the solar system and/or beyond? If some exoplanets are hospitable to life, how could we identify those rare pearls and how could we find evidence for eventual forms of life? This is the Holy Grail for the whole scientific community and well beyond, a major challenge for the coming century.

The Definition of the Planets by the IAU
General Assembly, Prague, 24 August 2006

Resolutions

Resolution 5A is the principal definition for the IAU usage of "planet" and related terms.

Resolution 6A creates for IAU usage a new class of objects, for which Pluto is the prototype. The IAU will set up a process to name these objects.

IAU Resolution: Definition of a "Planet" in the Solar System

Contemporary observations are changing our understanding of planetary systems, and it is important that our nomenclature for objects reflect our current understanding. This applies, in particular, to the designation "planets". The word "planet" originally described "wanderers" that were known only as moving lights in the sky. Recent discoveries lead us to create a new definition, which we can make using currently available scientific information.

Resolution 5A

The IAU therefore resolves that planets and other bodies in our solar system, except satellites, be defined into three distinct categories in the following way:

(1) A "planet" is a celestial body that (a) is in orbit around the Sun, (b) has sufficient mass for its self-gravity to overcome rigid body forces so that it assumes a hydrostatic equilibrium (nearly round) shape, and (c) has cleared the neighbourhood around its orbit.

(2) A "dwarf planet" is a celestial body that (a) is in orbit around the Sun, (b) has sufficient mass for its self-gravity to overcome rigid body forces so that it assumes a hydrostatic equilibrium (nearly round) shape, (c) has not cleared the neighbourhood around its orbit, and (d) is not a satellite.

(3) All other objects, except satellites, orbiting the Sun shall be referred to collectively as "Small Solar System Bodies".

IAU Resolution: Pluto

Resolution 6A

The IAU further resolves:

Pluto is a "dwarf planet" by the above definition and is recognized as the prototype of a new category of trans-Neptunian objects.

According to the definitions above, the solar system has eight planets: Mercury, Venus, Earth, Mars, Jupiter, Saturn, Uranus and Neptune. The list of dwarf planets currently has four members: the largest asteroid, Ceres, and three trans-Neptunian objects: Pluto, Eris and Makemake.

See also

http://www.iau.org/public_press/news/detail/iau0603/

How to Explore Planets?

Although the motions of planets — those "wandering objects" — have been known since antiquity, their exploration as physical objects started only at the beginning of the XVIIth century, with Galileo and his new refractor.

1.1 The Earth in space

Among the many consequences of the advent of the space era, in the second half of the XXth century, the very concept of our own planet has been questioned. With the first images of our planet as seen from space, the Earth has appeared as a solar system planet; these images could be compared with those of other telluric planets, then those of giant planets, also acquired by spacecraft in the 1970s and 1980s. This was the beginning of comparative planetology, which aims at studying all planets globally, in their differences and similarities.

Since the Copernican revolution, we have known that the Earth rotates around the Sun. In antiquity, with the exception of a few precursors like Aristarchus of Samos (310–230 B.C.), philosophers and scientists, in particular Aristotle, put the Earth at the center of the Universe. In order to account for the peculiar motion of planets on the celestial sphere, they needed a complex geometrical system based on circular motions. The final stage of this concept was Ptolemy's model, two centuries After

Christ, which used a combination of deferents and epicycles to describe the irregular motions of Mercury, Venus and Mars. This system lasted until the Copernican revolution.

Following the ideas proposed a century earlier by another precursor, Nicolas of Cusa, Nicolas Copernicus (1473–1543) stated the founding principles of the heliocentric system in a famous posthumous book: planets all orbit the Sun in the same direction; the dimensions of the solar system are negligible compared to the distance between the Sun and the nearest star. Copernicus' theory, first badly received because of its opposition with the religious dogma which prevailed at that time, was finally recognized thanks to the subsequent works of Kepler (1571–1630), Galileo (1564–1642) and then Newton (1642–1727). Following the observations of his teacher, the Danish astronomer Tycho Brahé, Kepler announced the three famous Kepler's laws, which describe the motion of planets as ellipses having the Sun as one focus. Galileo, the first scientist to use an astronomical refractor, opened the era of astronomical observation. He discovered, in particular, the lunar craters, the phases of Venus, and the main satellites around Jupiter, now called the Galilean moons. Finally Isaac Newton, stating the laws of universal gravitation, definitely confirmed the heliocentric system by giving it the theoretical basis which was previously missing.

In addition to the Earth, the solar system had then five planets, all visible with naked eye: the three telluric ones, Mercury, Venus and Mars, and the two giant ones, Jupiter and Saturn. Two other giant planets were later discovered. The discovery of Uranus belongs to William Herschel (1738–1822) who built a "large" telescope with a 15-cm aperture, significantly improving the detection limit of celestial bodies. The second discovery illustrates the growing success of celestial mechanics, capable of calculating planetary orbits with an increased accuracy, by taking into account gravitational perturbations from neighboring bodies. From a precise analysis of Uranus' trajectory, it appeared that the orbit had to be perturbed by the presence of an unknown body, farther from the Sun. In 1846, two astronomers working independently, John Couch Adams in England and Urbain Le Verrier in France, simultaneously determined the position of the eight planet, which was immediately discovered at the position announced by Le Verrier.

We now had eight planets in the solar system. Still, this was not the end of the story. For decades following Neptune's discovery, astronomers searched for a ninth planet: its presence was suspected on the basis of anomalies observed in the orbits of Uranus and Neptune. These anomalies were later explained by the uncertainties associated with these calculations. Anyway, the hunt led, in 1930, to the detection of Pluto, immediately called the ninth planet. But this planet was not massive enough to explain the so-called anomalies of Uranus and Neptune' trajectories; the hunt for "Planet X" went on. In the 1940s, Kenneth Edgeworth and later Gerard Kuiper suggested the existence of a population of small objects located beyond Neptune's orbit; their argument is based on the distribution of matter in the protoplanetary disk, which should naturally continue beyond 30 AU. We had to wait until 1992 for the first transneptunian object to be discovered. It then appeared that Pluto is one of the largest representatives of this family. The International Astronomical Union officially endorsed this statement in 2006 by removing Pluto from the list of planets. Our planetary system is thus complete: relative to its heliocentric distance, the Earth is the third one of the list; it is also the largest and the most massive telluric planet.

1.2 Telescopic observations

1.2.1 Draw me a planet...

This is how astronomical observation starts: observers draw, as accurately as they can, what they discover at the eyepiece of their instrument. In particular, with his refractor, Galileo observed, without being able to explain them, the "handles" which surround Saturn and their temporal variations. The explanation of this phenomenon was given in 1659 by Christiaan Huygens: what is seen is a thin ring system, observed from the Earth under a variable inclination. Jean-Dominique Cassini, founder and first director of Paris Observatory, discovered the division now named after him. His drawings of Jupiter are surprisingly precise: the structure in zones and belts are shown, as well as the Great Red Spot, still present over three centuries after; Cassini also deduced the rotation period of Jupiter, surprisingly short for a body of such size (less than ten hours).

For over three centuries, the cartography of planets has been based on visual observations and drawings, with obvious success. Still, one has to be aware of the limitations of the method, associated with possible optical illusions which may affect visual observations; a clear illustration of this problem is given by the so-called discovery of channels on Mars. In the 1870s, the Italian astronomer Schiaparelli reported the discovery of linear structures on the surface of Mars, called "canali". At that time, the Martian atmosphere was known to be very dry, and some astronomers saw in these channels the sign of extraterrestrial life... A wide controversy followed among the astronomical community. The end of the myth only came in the 1960s, when the first spacecraft demonstrated that the "canali" were no more than an optical illusion.

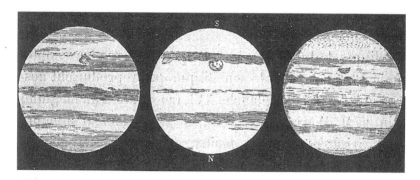

Figure 1.1. The first drawings of Jupiter by Jean-Dominique Cassini, the first Director of Paris Observatory. The band/zone structures are identified, as well as the Great Red Spot (center, top). (© L. Selva, Tapabor).

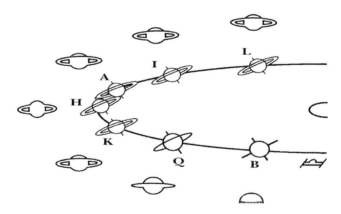

Figure 1.2. This drawing illustrates the first correct interpretation of the nature of Saturn's rings, made by Christiaan Huygens in 1659 in "Systema Saturnum." He shows that the "handles" observed by Galileo are a very thin ring, seen from the Earth at an angle which varies with time, depending on the position of the Earth relative to the plane of the rings. According to C. Sagan, Cosmos, Mazarine, Paris, 1981.

1.2.2 From photographic plate to numerical camera

The first photographic images of planets were obtained at the end of the XIXth century. Still, this technique generalized only around the 1950s. With respect to visual observations, this technique has a great advantage, as it removes possible subjectivity. The photographic plate, then the film, was the main tool of astronomers until the advent of the numerical camera at the beginning of the 1980s. The use of Charged Couple Devices (CCDs) is a real revolution for astronomy, comparable to the use, in daily life, of digital cameras with respect to traditional photographic cameras. Numerical cameras are more reliable, simpler to use, and with an increased dynamical range; even more important, numerical data can be directly used for data processing and analysis. The rediscovery of Comet Halley at Mount Palomar, in 1982, at less than 10 arcsec from its predicted position, is among the first astronomical successes of the CCD; four years before perihelion, the comet was then at 11 AU from the Sun.

Planetary imaging is essential for studying the morphology of planetary disks and their temporal evolution, as well as for an accurate determination of the orbits of small bodies (comets, satellites, asteroids). In addition, the photographic plate and then the CCD camera give access to another type of measurement: photometry, which allows us to determine accurately the amount of light received at each point of the image — something the eye cannot do. An interesting application of this technique is the observation of stellar occultations, when by chance a solar system object transits in front of a star. If the planet has an atmosphere, the continuous decrease of the stellar flux at the beginning of transit, then its increase after the transit, gives us information about the atmospheric properties of the planet. In the case of Uranus and Neptune, this type of observation has allowed the detection of a ring system, later confirmed by the Voyager 2 spacecraft and the Hubble telescope.

1.2.3 From high energies to the radio waves

How can we determine the chemical nature of the planetary atmospheres and surfaces? The images, as beautiful as they can be, do not give the answer. Another tool is needed: the spectroscopy, i.e. the analysis of the planetary radiation as a function of its wavelength. We know that visible light can be decomposed, with a prism for instance, into the colors of a rainbow. In the same way, it is possible to analyze the radiation, invisible to the eye, which comes at shorter wavelengths or higher frequencies (these are the ultraviolet, X- and gamma rays) and at longer wavelengths or lower frequencies (these are the infrared , submillimeter, millimeter and radio waves). For each spectral domain, one needs specialized instrumentation including the detector, which captures the signal, and a spectrometer, which disperses it as a function of the wavelength. Modern astronomical instruments now combine the imaging and spectroscopic capabilities in a single device: using two-dimensional detectors, they allow us, using a push-broom technique, to build three-dimensional maps (two spatial dimensions and one spectral dimension). The advantage of this method is that, in many cases, observation at different wavelengths allows us to probe, in an atmosphere, several layers of different altitudes. Ultraviolet radiation, capable of dissociating and ionizing molecules, allows us to probe the high altitude atmospheric levels (their stratosphere and thermosphere), while infrared radiation, of less energy, typically probes the lower levels (their troposphere). In this way, 3-D cartography can be achieved. Another major advantage of spectroscopy is the identification of atmospheric constituents or minerals on a surface. Mineralogical maps, or maps of specific atmospheric constituents, can be retrieved, which are precious tools for our understanding of the planet's climate and evolution.

1.2.4 Observing from Earth

Naturally, the progresses achieved in astronomical observation are directly linked to the size increase of the telescopes, whose primary mirrors now reach the 10-m range. Using such a telescope has two advantages: first, the collected flux is larger and the measurement is thus more sensitive; second, its diffraction limit (the smallest detectable angular distance on the sky) is smaller, which allows us to observe smaller details. As an example, a 2-m diameter telescope has, in the visible range, a diffraction limit lower than a tenth of an arc second. The mean angular diameters of planets are respectively 40 arcsec and 15 arcsec for Jupiter and Saturn. In the case of Venus and Mars, they strongly vary as a function of their distances to the Earth; their maximum values are about 60 and 20 arcsec respectively. Thus, high-precision images could in principle be achieved by a 2-m telescope. Unfortunately, the turbulence of the terrestrial atmosphere strongly limits the image quality. In the case of a stable atmosphere (which is found in the best astronomical sites) , this limit (called the seeing) is typically 0.5 to 1 arcsec. The image quality is thus degraded by a large factor. To compensate for this defect, astronomers have designed a correction method called adaptive optics. They use a reference star to monitor the atmospheric turbulence (by measuring the distortion of the wave-front) and they correct it by adjusting, in a servo-control loop, the motions of the actuators of a deformable mirror. This method is now generally used on all large telescopes. Reference stars may by natural stars in the field of view or laser guide stars.

1.3   Observations from space

1.3.1 Observing from Earth orbit

Here is a first reason for the astronomers to go to space: to get rid of the atmospheric turbulence. This was the main reason for the Hubble Space

Figure 1.3.   Planet Mars observed by the camera of the Hubble Space Telescope (© NASA).

Telescope (HST), launched in 1989, to operate in Earth orbit. Its camera has given us planetary images of exquisite quality. Apart from image quality, the HST has another good reason to operate outside the terrestrial atmosphere: by doing so, it gets access to new spectral ranges invisible from Earth, first the ultraviolet, then the infrared range as new instruments were implemented.

The spectral range extension is, in itself, a sufficient reason for astronomers to go to space (Fig. 1.4). Indeed, certain gases in the terrestrial atmosphere make it opaque at some wavelengths. This is particularly true for water vapor, but also, to a lower extent, for carbon dioxide, methane, ozone... There is one clean window, the visible range, which extends from about 0.4 to 0.8 micrometers; this is the range where the human eye is optimized. At shorter wavelengths, in the higher energy range, the atmosphere is completely opaque to UV, X- and gamma radiation. UV radiation is absorbed by the ozone layer, which thus prevents the destruction of living organisms on the continents. In the infrared range, a few spectral windows can be used for ground-based observations; it is also the case at millimeter and radio wavelengths. However, even in these windows, it is very difficult to search for planetary atmospheric signatures also present in the terrestrial atmosphere.

The infrared range is very well suited for the study of neutral planetary atmospheres, because the molecules present in these atmospheres have strong spectral signatures, associated with their motions of rotation

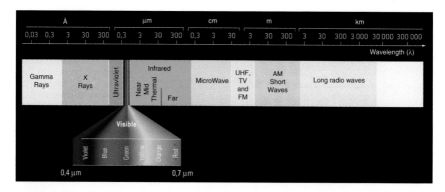

Figure 1.4. The dispersion of light in the electromagnetic spectrum (© T. Encrenaz, Searching for water in the Universe, Belin 2004).

or vibration. In addition, the planetary flux is maximal in this spectral range. This is why planetology has strongly benefited from the ISO (Infrared Space Observatory) mission, launched by the European Space Agency (ESA) in 1995, in operation in Earth orbit until 1998. As will be seen later, the ISO infrared spectrometers have made major discoveries regarding the atmospheres of the giant planets.

1.3.2 Coming close to a planet...

For planetologists, Earth-orbit observations also bring great advantages compared to ground-based observations. Still, planets are very distant... For an in-depth exploration of a planet, the best method is approaching it. This is the whole motivation of planetary space exploration since its beginning some fifty years ago. Let us put aside the manned Moon exploration whose objectives were not mainly driven by science (although science has been able to extract the best profit, in particular for solar system dating). Planetary exploration has developed following several milestones: first the flybys, then the orbital missions, then the landers which came at the surface and finally the rovers which were able to move on it. This is where we are now. The next step (which has already started in the case of comets and asteroids) will be the return of planetary samples, possibly from Mars, which could take place around 2025.

Let us be very clear: we are talking here about robotic exploration. Some people think that, in a more distant future, manned exploration will take place again, on Mars or elsewhere. There are multiple motivations for such an enterprise, but it is important to stress that the scientific return is far from being a major justification. Let us take the case of the lunar samples: Soviet landers brought samples at the same time as the Apollo spacecraft (although in a much smaller quantity); thus, putting a man on the Moon was not a mandatory condition for achieving the scientific objectives of the mission. On solar system planets, robotic missions have achieved incredible results, made possible by high technology and miniaturization.

Initiated in the 1960s, the exploration of telluric planets has been a long adventure, full of obstacles. In the case of Mars in particular, multiple failures have occurred — more than 50% of all missions launched toward the red planet have failed. In the beginning, many spacecrafts were lost at

Figure 1.5. The Very Large Telescope of the European Southern Observatory at Cerro Paranal (Chile) (© ESO).

Figure 1.6. Planet Saturn observed with the adaptive optics instrument NAOS-CONICA at the VLT, ESO (Chile) (© ESO).

launch, or missed their target, or crashed at the surface. Planetary exploration has been a school of patience! In spite of all these difficulties, spectacular successes have been achieved. In the case of Mars, the main milestones have been in the 1970s, with Mariner 9 and especially the Viking mission. From this perspective, the Viking mission, composed of two identical orbiter and landers, appears as an exceptional technological performance; it was also a tremendous scientific success, and its database is still used today as a reference by the scientific community. The only bad news was that Viking did not discover any trace of life on Mars, and this search was its major motivation... As a result, the exploration of Mars was suspended for almost twenty years! It started again at the end of the XXth century with, on the NASA side, the Pathfinder rover, followed by Phoenix, and a series of orbiters (Mars Global Surveyor, Mars Odyssey, Mars Reconnaissance Orbiter) and rovers (Spirit and Opportunity, and now Curiosity). ESA started its planetary exploration program in 1986 with the Giotto spacecraft toward Halley's comet, and realized a success-fulcome back in 2003 with the Mars Express orbiter. Other projects are in preparation to pursue the exploration of Mars, including the MAVEN and ExoMars-2016 orbiters and the ExoMars-2018 rover.

In the 1980s, planet Venus was the privileged target of the Soviet space exploration with the Venera missions, equipped with orbiters and landers. However, the very high temperature and pressure at its surface make it very difficult for a probe to survive. After several failures, the first images of the surface were returned to Earth by Venera 13 in 1982. In parallel, NASA explored Venus with the Pioneer Venus mission, then by the Magellan orbiter which obtained, with its radar, a complete map of the surface in the early 1990s. In 2006, ESA launched an atmospheric probe, Venus Express, still in operation today. Mercury, although more difficult to approach due to its vicinity to the Sun, was not forbidden: the planet was already studied by NASA in the 1970s during several flybys of the Mariner 10 spacecraft; the exploration of Mercury has started again in 2008 with the Messenger spacecraft from NASA, and is expected to continue around 2020 with the Bepi Colombo mission of ESA.

Regarding the giant planets, all the NASA missions, including the first ones, have been successful: launched in the 1970s, Pioneer 10 and 11, followed by Voyager 1, explored Jupiter and Saturn, while Voyager

2 made successive flybys of all four giant planets between 1979 and 1989; the Voyager missions have been an outstanding success and their results are still used as a reference today. In the case of Uranus and Neptune, they will stay so for at least a decade, as no future mission is presently planned toward these planets. With the Galileo mission, launched in 1989 toward the Jupiter system, NASA entered a new type of long-term planetary monitoring, with an orbiter, in operation from 1995 until 2003, and a descent probe, which entered the Jovian atmosphere and successfully returned data to Earth down to a pressure over 20 bars. The last chapter of this saga, the Cassini–Huygens mission, jointly led by NASA and ESA, is the most brilliant and successful example of internal cooperation for planetary exploration. The Cassini spacecraft has been exploring the Saturnian system since 2004, while the Huygens probed successfully landed on Titan's surface in 2005. Another ambitious project, called JUICE, has been selected by ESA for exploring again the Jovian system, with a launch planned around 2022.

Figure 1.7. The descent module of the Viking mission. Two identical modules operated on the Martian surface for a period of several years. The Viking mission was a great scientific and technological success (© NASA).

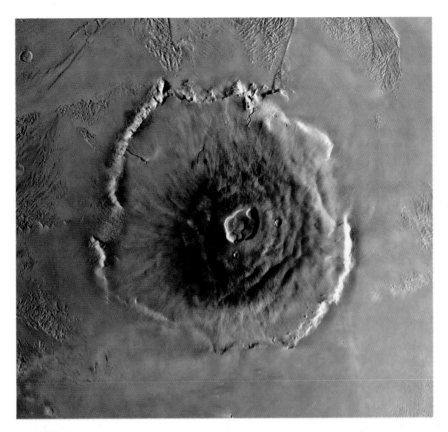

Figure 1.8. The Martian volcano Olympus Mons photographed by the Viking orbiter. With an altitude over 20 kilometers, it is the highest known mountain in the solar system (© NASA).

What have we learned from planetary space exploration? A lot regarding the surfaces, the composition and structure of the atmospheres, the magnetospheres ... all that based on two types of instruments. For flybys and in-orbit observations, spacecraft are equipped with remote sensing instruments; as for ground-based observations (but obviously the instruments have to be simple, robust and light): cameras, spectrometers operating at different wavelengths, from gamma rays to the radio range. Descent probes, landers and rovers are equipped with *in situ* instruments for collecting and analyzing samples, gaseous or solid. They include chromatographs for the study of clouds and mass spectrometers for neutral or ionized species. The instrumental payload also includes instruments for plasma physics: radio and plasma receivers for analyzing radio waves, particle analyzers, magnetometers and other more specific instruments. A requirement common to all is reliability: once a spacecraft has been launched, there is no possible contact but telemetry. Another strong constraint is the mass, which must be kept as low as possible for cost reasons. Multiples efforts have been made to miniaturize space instrumentation. After several decades of planetary exploration, a conclusion can be drawn: space instruments are extraordinarily reliable. It may happen, unfortunately, that a spacecraft is lost at launch or, less often, during an orbit maneuver, but the failure of a space instrument is exceptional.

After this short overview of planetary space exploration, one question must be raised: after all these space missions, are ground-based observations still worthwhile? The answer is definitely yes, for several reasons. The first one is that the number of planetary space missions remains limited: as an example, there are presently no prospects for a new space exploration mission to Uranus and Neptune. The second reason is that space instrumentation must be simple, because it must be light and

Figure 1.9. Io, the Galilean satellite closest to Jupiter, observed by the camera of the Galileo spacecraft (© NASA).

Chapter 1. How to Explore Planets?

reliable; it has to be based of already proven technologies. Ground-based instruments, in contrast, are much more sophisticated and benefit from the latest technological improvements. Finally, ground-based observations allow astronomers to monitor continuously time-variable phenomena, and also to obtain global instantaneous images of the planets — something in-orbit space missions cannot do because they are too close to the planet. Ground-based and space planetary observations are thus fully complementary, which is true also for astronomical observations at large. Planetology makes the best use of large ground-based facilities, like the Very Large Telescope in the optical and infrared range, and the ALMA large array in the submillimeter/millimeter range.

1.4 Searching for exoplanets

Are there, outside the solar system, other planets comparable to ours, and could they host life? This question has been raised by humanity since its early origins, and the recent discoveries achieved at the end of the XXth century have only confirmed its pertinence: the Sun is a very common star in our Galaxy which hosts over a hundred billion of them. Why would the solar system be unique? Over the last decades, many observing programs have been designed and implemented to address this question. Astronomers soon realized that a direct detection was extremely difficult, because the planetary radiation is lost inside the radiation of its host star. Astronomers have thus tried an indirect method, to determine by astrometry the motion of the star with respect to the center of gravity of the star–planet system. This method was successfully used for detecting companion stars; this is how the German astronomer Friedrich Bessel was the first one, in the XIXth century, to detect a companion around Sirius. Unfortunately, in the XXth century, astrometry techniques were not sensitive enough for detecting planets (see Chapter 6, Section 1.1).

In the meantime, significant progress has been made in our knowledge of the first phases of stellar and planetary formation. We now know that stars form after the collapse of a fragment of molecular cloud within a disk. At the center, the matter concentrates to form the protostar, while planets form within the disk from the accretion of solid particles (see

Figure 1.10. Titan's surface, as seen by the DISR instrument aboard the Huygens probe. The probe landed successfully on January 14, 2005 and sent to Earth the first images of Saturn's biggest satellite (© ESA).

Chapter 6, Section 3). This is one more reason to search for exoplanets... Following the first detection of a disk around the Vega star by the IRAS satellite (InfraRed astronomical Satellite) in 1983, astronomers have accumulated observations of disk around young stars, either very young protoplanetary disks or older debris disks (see Chapter 2, Section 2). Infrared and millimeter observations, using either ground-based millimeter telescopes or the Earth-orbiting infrared satellite Spitzer, have allowed astronomers to understand the physical and geometrical properties of these disks, and also to determine their ages.

In 1992 came a surprise: the polish astronomer Alexander Wolszsczan announced the discovery of two planets around a very special star, the pulsar 1257 + 12. Pulsars are peculiar stars at the very end of their lives. Following an implosion phase, they have become neutron stars, with an extremely high density, rotating extremely fast. They emit a periodic radio signal, modulated by their rotation period. If a planet is present, this periodic signal shows a detectable perturbation (see Chapter 6, Section 1.2). These two planets are the first two exoplanets ever discovered; however it is hard to imagine that life could exist in such extreme conditions.

Other observing efforts, in contrast, have focused on solar-type stars. The method used this time is velocimetry, i.e. the measurement of the relative velocity of the host star: if its motion is perturbed by the presence of the planet, it shows a modulation, which can be detected from Earth. The velocimetry method is based on the measurement of the Doppler shift (i.e. the spectral shift induced by the velocity of the star) in the spectrum of the star. The objective is the same as for astrometry: to measure, in the motion of the star, a periodic modulation due to the presence of a planet, but the velocimetry method is much easier to achieve than astrometry. After several years of continuous tracking, in August 1995, the announcement of the discovery of a planet around a solar-type star exploded like a bomb: Michel Mayor and Didier Queloz, from the Observatory of Geneva, made this discovery at the Haute Provence Observatory, using a 1.93 cm telescope and a high-resolution spectrometer.

For a whole decade, velocimetry was the privileged method for detecting exoplanets. It can be achieved from the ground using telescopes of modest size (2–4 m); what is needed is a very precise spectrometer, operating in vacuum, allowing very stable measurements. The more massive the planet, the larger the motion perturbation and the easier the detection. As an example, the presence of Jupiter induces a periodic motion of 12 m/s on the solar motion; now the most precise instruments reach the limit of 1 m/s. As of the end of 2012, over 800 exoplanets have been discovered, and most of them have been detected by velocimetry.

Other complementary methods have also been developed. The most successful is the observation of planetary transits (see Chapter 6, Section 1.4). The method consists in measuring very accurately the flux of the host star as a function of time. If it happens that the orbit plane of a planet around the star also contains the observer's position, then the planet passes in front of the star (this is called a transit) and the stellar flux is partially occulted during the transit. This observation, repeated along the transits, allows one to detect the presence of the planet unambiguously, and also to determine its radius. Combined with the velocimetry technique, it allows the determination of the mass and density of the planet. A Jupiter-like planet, ten times smaller than its host-star, transiting in front of the Sun, induces a 1% decrease of the solar flux; such a variation is observable from Earth with high-precision photometry. In contrast, an Earth-like transit induces a stellar flux decrease of only 10^{-4}, too weak to be monitored from the ground; the measurement requires a stability over time which can be reached only from space. This is why, in December 2006, the space mission CoRoT was launched by the French space agency CNES, with the objective of detecting rocky planets (also called "super-Earths"). Several tens of exoplanets have been detected by CoRoT and later confirmed by velocimetry; indeed, the transit method can lead to the detection of "false positives" induced by the variability of the star, and a confirmation of the exoplanet's nature by velocimetry is needed. In March 2009, NASA launched Kepler, a more sensitive satellite with the same objective. Kepler has been extraordinarily successful: in addition to the few tens of exoplanets firmly identified, Kepler has announced the detection of over two thousand potential candidates. Their host stars are often too faint to allow the confirmation of the exoplanet by velocimetry, but the results can be used nevertheless for statistical studies. The detected exoplanets show an extreme variety in their orbits and physical parameters. Obviously, the exploration of exoplanets is exploding and many new discoveries are in front of us...

A decade ago, direct imaging of exoplanets looked beyond the scope of ground-based observation. Still it has shown to be feasible, provided the astronomers consider, instead of solar-type stars, M-dwarfs. Those stars are much fainter than the Sun, and the planet/star flux contrast is thus consequently enhanced (see Chapter 6, Section 2.8.5). Several exoplanets have been detected with this method. Finally, there is another original method, called "gravitational microlensing". This method consists in a systematic search for occultation events which occur when a star–planet system transits in front of a distant object. This method of gravitational lensing was originally designed to search for dark matter in the Galaxy. During the occultation, the flux of the distant object is magnified by a lensing effect and, if a planet is present, the light curve shows a character-istic signature. The microlensing technique has led to the detection of a few exoplanets, including a very small one.

In the forefront of this exploding research field, several instrumental projects are developing. Instruments for direct imaging are developed for large ground-based telescopes, like the SPHERE instrument on the Very Large Telescope of ESO. More sophisticated instruments are in preparation for the next generation, the Extremely Large Telescope, whose diameter will reach about 40 m. Space projects are also under study, on the ESA and NASA sides. Infrared coronagraphic imaging and spectroscopy are emphasized, to get the best planet/star flux contrast. The JWST, successor of the HST, will be an important tool for this study; other dedicated space projects are also under study. Definitely, the exploration and characteriza-tion of exoplanets, together with planetary exploration and cosmology, will be one of the major challenges of astronomy in the coming century.

The Birth of Planets

How did planet Earth form, and what is its age? The question has been raised since antiquity and all mythologies, including the Bible, have offered possible solutions. An age has indeed been settled on by Church theologians who found that the Earth was a few thousand years old. With the advent of the heliocentric system, the question has naturally evolved towards the origin and dating of the whole solar system.

2.1 A formation within a disk

At the end of the XVIIth century, the orbits and motions of the six planets identified at that time — Uranus and Neptune were still missing — were well known. An observational fact is obvious: all planets rotate around the Sun in the same direction, and their orbits are nearly circular, coplanar and concentric. This simple statement is the basis of the premonitory intuition of the German philosopher Emmanuel Kant (1724–1804), later followed by the French physicist Pierre Simon de Laplace (1749–1827): the solar system was born from a rotating nebula which collapsed into a disk. At the center, the matter concentrated to form the proto-Sun. Within the disk, due to some inhomogeneities, solid particles accreted locally to form planetesimals, then bigger objects and finally a few planets.

This simple scenario, first proposed more than three centuries ago, is surprisingly modern. Indeed, it describes in its main characteristics the scenario generally accepted today for the formation of the solar system. This scenario is based on simple observations: the planets' orbits are all close to the plane of the Earth orbit, called the ecliptic. At the time of Kant and Laplace, there was no mathematical nor physical validation of this model, which was not unanimously accepted first. Other models were developed, including the vortex theory of René Descartes at the XVIIth century, and the theory of a passing-by star (or even a comet) that snatched a small fraction of the solar matter to make the planets. At that time the true nature of comets was not known: it was revealed only later by the English astronomer Edmund Halley, in 1949. Until the 1950s, the model of the primordial nebula was controversial. A peculiar question is a matter of debate. The rotation period of the Sun is 27 days while, according to Laplace' model, by conservation of the angular momentum, it should be much less. In the solar system, about 99% of the angular momentum is in the planets, especially Jupiter. How did the Sun transfer its momentum to the planets? The primordial nebula model described by Laplace does not give an answer to this question. We now know that the transfer of angular momentum had been achieved through the interaction with the very turbulent protoplanetary disk, and also through the solar magnetic field and the ejection of solar matter in the form of solar wind.

2.2 A common scenario in the Universe

Since the 1980s, observations have revealed a new phenomenon: the existence of disks around young stars. Before, circumstellar disks were known only around evolved stars. In 1983, the IRAS satellite (InfraRed Astronomical Satellite) discovered an infrared excess around the young star Vega. As the infrared radiation is a tracer of cold matter, this result was soon interpreted as the presence of a dusty disk around the star. The A-type Vega is not really a young star, but it is younger than the Sun. Its disk is not a protoplanetary disk but an older one, so-called "debris disk". It is now believed that this debris disk is the residual left by the star after its accretion phase or by planets after their formation. With the development

Figure 2.1. The infrared satellite IRAS (InfraRed Astronomical Satellite). Launched in 1983, the satellite has conducted, from Earth's orbit, a deep survey of the sky in four wavelengths of the mid and far infrared (12, 25, 60 and 100 μm). (© NASA/NIVR/SERC).

of infrared astronomy, in particular with the ISO and Spitzer satellites, many other examples of debris disks have been discovered.

Another decisive milestone was, in 1984, the discovery of a disk around the star Beta Pictoris, by ground-based coronagraphy in the visible range. This is again a debris disk, but this time much younger than Vega's. About this star, a very surprising result was found by a group of scientists led by A.Vidal-Madjar: in the spectrum of the star, transient phenomena are observed, interpreted as the signature of cometary-type objects falling on the star. More recently, a planet has been discovered in orbit around this star. The analogy with a planetary system gets more precise.

In parallel with these discoveries, the scenario of stellar formation developed more and more, thanks to the study of young objects at different stages of their evolution: Herbig–Haro objects, characterized by the emission of a strong bipolar flux; T-Tauri and Fu-Orionis stars, characterized by very violent stellar winds. HST observations show evidence for protoplanetary disks; their properties are studied by ground-based millimeter telescopes and by the infrared satellite Spitzer. Complementary to these observations, numerical simulation models of the accretion phenomenon allow astronomers to understand the early stages of stellar formation: the process starts, within the interstellar medium, with the fractionation of a cold molecular cloud whose density and rotation increase as its contraction accelerates. Beyond a critical mass, the cloud collapses into a disk perpendicular to its rotation axis: the scenario is identical to the one proposed by Kant and Laplace. Current models now take into account the magnetic field of the protostar and the strong turbulence within the protoplanetary disk; they can now account for the whole sequence of observations.

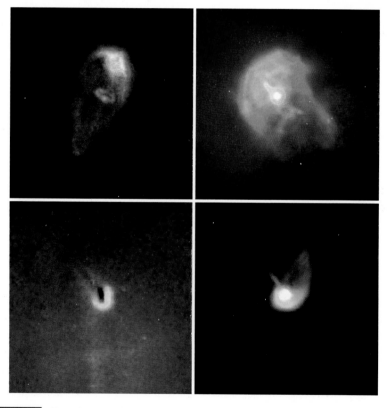

Figure 2.2. Examples of protoplanetary disks observed by the camera of the Hubble Space Telescope (© NASA).

Figure 2.3. A protoplanetary disk observed with the HST: the Herbig–Haro object HH30. The disk appears dark, in absorption in front of the light of the young star. Perpendicular to the plane of the disk, a violent bipolar jet is shown (here in green) (© NASA).

Nowadays, astronomers have come to the conclusion that the stellar formation process is commonly driven by the collapse of a rotating molecular cloud. Planets, if they exist, must be formed within the disk, following some mechanisms that remain to be better understood. One observational fact is accepted: the lifetime of the protoplanetary disk is very short, 10 million years at maximum; it is also the age of T-Tauri stars. These stars, through their very strong stellar winds, make the disk disappear by expelling its gas and dust components. If planets can form in the disk, they must do so in a very short time, in less than 10 million years or so. In this scenario, what are the debris disks? They could be the equivalent of the Kuiper Belt in our solar system. Debris disks, however, are significantly more massive than the Kuiper Belt. There is a possible explanation to this difference: most of the content of the Kuiper Belt might have been ejected out of the solar system, due to some migration of the giant planets in the early stages of the solar system's history; we will come back later to this scenario based upon numerical simulations (see Chapter 5, Section 1.3).

Figure 2.4. Debris disk around the star AU Mic (© P. Kalas, 2004).

2.3 What is the age of the solar system?

For understanding the origin and history of solar system objects, a chronology is essential. We have known it since the 1970s, thanks to the analysis of lunar samples brought back by the Apollo astronauts and the robotic Soviet missions.

2.3.1 The chronology of lunar samples

How can we date a rock sample, either terrestrial or extraterrestrial? The method consists in analyzing the abundances of radiogenic species, compared to those of their stable isotope. Several pairs of elements can be used, with lifetimes ranging from a billion years to a million years or even shorter in some cases. For solar system dating, the rubidium–strontium couple is used, with a lifetime of 47 billion years. The analysis of lunar samples has demonstrated that the Moon, like the Earth and the parent bodies of meteorites, were formed simultaneously, some 4.56 billion years ago.

2.3.2 What is the age of the Sun?

This age is also the age of the Sun, a G-type star which has now reached the middle of its lifetime. For many decades, stellar physics, using the systematic study of stars of different ages, has been able to reconstruct the life and death scenarios of stars with a great precision. The main parameter in this scenario is the mass of a star. The more massive the star, the shorter its life. O and B stars are the most massive ones, then A stars like Vega. The Sun, a "mean" G star, has an expected lifetime of ten billion years. M-dwarfs have a much longer lifetime. The evolution of the different types of stars is represented in the famous so-called Hertzprung–Russell diagram (Fig. 2.5) which describes the absolute luminosity of the stars as a function of their temperature (i.e.

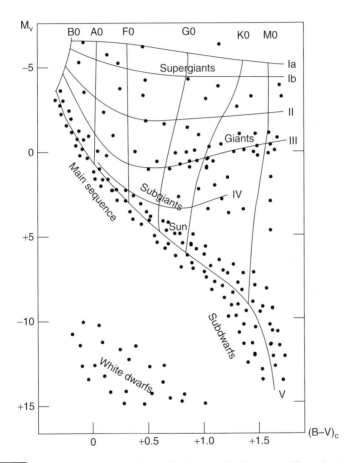

Figure 2.5. The Herzprung–Russell diagram for the classification of stars. The ordinate axis indicates the magnitude M_v of the star. The amount (B–V), shown in abscissa, indicates the difference in magnitude of the star in the (B) and (V) bands of the optical spectrum; (B–V) is a decreasing function of temperature.

their spectral type). The Sun is presently in the "main sequence", a phase corresponding to the burning of hydrogen into helium, which will continue for a few more billion years. At the end of this process, in the case of massive stars, the nucleosynthesis will continue to form heavier elements: carbon, nitrogen and oxygen. The Sun itself will become a red giant. Its radius will inflate up to reach the inner planets... The Earth will of course vaporize in the process, as well as all other planets and, at larger distances from the Sun, the Kuiper Belt. The sublimation of the Kuiper Belt is expected to produce a strong emission of water vapor, since water (presently in the form of ices) is the main component of small bodies in the outer solar system objects. Actually, such a water emission may have been detected from a few evolved stars, which might be the signature of this phenomenon. The Sun will end its life as a white dwarf, a very small star with an extremely high density. This is our expected fate... in about five billion years from now!

2.3.3 Surface dating by crater counting

There is another method for dating the surfaces of solar system bodies. It consists in counting the density of impact craters of meteoritic origin on their surface. The bare bodies like the Moon, Mercury or moons of the giant planets, are easier to date as the impact craters have not been altered since their formation. Mars, whose atmosphere is very tenuous, is also a good target for this method. The older a surface is, the more heavily cratered it is. Thanks to the absolute dating achieved for the Moon surface from the analysis of lunar samples, it is now possible to plot the curve indicating the age of a surface as a function of its impact crater rate. This curve illustrates that, at the beginning of the solar system's history, the meteoritic bombardment was much heavier than today. This is in agreement with what we know about the highly energetic activity of the young star and the turbulent state of the early disk. More precisely, it appears that a maximum occurred in the impact cratering rate about 800 million years after the planets' formation; we have no information about what might have happened before this date. This peak in the impact cratering rate, called the Late Heavy Bombardment (LHB), also finds its explanation in the dynamical simulation models.

Anyway, the crater counting method allows us to determine the surface ages. We know that those of the Moon and Mercury are very old, which

Figure 2.6. An example of a very ancient surface: the heavily cratered terrain of Mercury, observed here with the camera of the Messenger probe (© NASA).

indicates the absence of internal activity or at least the fact that it stopped very early. The surface of Venus, in contrast, has a very low cratering rate, which shows that it has been remodeled within the last five hundred million years or so. Mars exhibits two very distinct hemispheres: in the northern part, the plains are mostly devoid of craters: they have been recently covered with volcanic flows. In the southern hemisphere, the highlands, more heavily cratered, are older. In the case of the Earth, the ocean floor is renewed by plate tectonics on a timescale of 200 million years. The continental crust, in contrast, may be as old as 4 billion years (see Chapter 3, Section 2).

2.4 The main steps of planetary formation

Let us go back to our scenario of planetary formation. The protosolar disk is mostly composed of gas, among which the most abundant one is hydrogen.

It also contains dust, including refractory particles (silicates, oxides, metals) and, at lower temperatures, ices (H_2O, NH_3, CH_4, CO_2 ...). At low temperature, the ices are more abundant than the refractory grains, made of heavier atoms (and thus less abundant, according to cosmic abundances). Solid particles tend to fall into the median plane of the disk, while the gaseous disk is thicker. All gaseous and dust particles rotate around the proto-Sun, and neighboring particles have comparable velocities: relative motions are thus

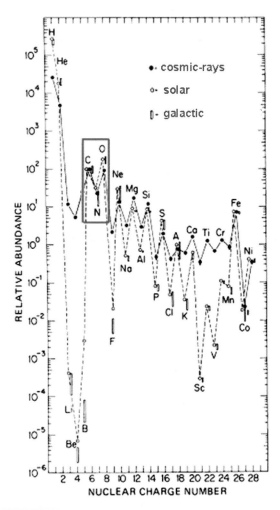

Figure 2.7. Cosmic abundances of the elements (© IRFU, CEA).

small. Particles frequently collide and stick together by electrostatic collisions; by this coagulation process, they can form millimeter to centimeter-sized grains. The observation of such particles collected in the stratosphere suggest that they may have a fractal structure. Following numerical simulations, this first phase could have been achieved in a few thousand years, a very short time with regard to planetary formation. In contrast, models have difficulties in reproducing the next phase allowing, from centimeter-sized embryos, the formation of kilometer-sized "planetesimals". The mechanism responsible for this process is still poorly understood. Whatever the reason is, some kilometer-sized planetesimals can later grow under the effect of gravitational interactions which favor, through collisions, the destruction or accretion of these bodies. Numerical simulations show that, at the end of this process, dominated by multiple collisions, a very small number of large bodies survive after a few million years. These protoplanets will again interact with each other, as demonstrated by the giant impact craters observed on the surface of Mercury and the Moon in particular. The solar system is the result of this gigantic pool game. We will see later that this scenario is also favored for the formation of exoplanets (see Chapter 6, Section 3.2).

2.5 Telluric planets and giant planets

As we have seen, planets form solid particles. Their growing rate and their composition is thus a function of the composition of available particles at the time of accretion. This composition critically depends on the temperature of the ambient medium.

In the protosolar gas, the relative abundances of the elements is driven by the cosmic abundances, found everywhere in the Universe. Hydrogen is the most abundant element, with a relative mass abundance of 75%. Helium comes next with slightly less than 25% by mass. Both were formed very early in the Big Bang model by primordial nucleosynthesis, and are in gaseous form in the protosolar disk. Heavier elements, starting with the most abundant ones, carbon, nitrogen and oxygen, represent altogether less than 2% of the total mass. These elements are associated with hydrogen to form small molecules like H_2O, CH_4, NH_3 ...; they are either in vapor or solid form depending on the temperature of their environment, which depends on their distance to the Sun. The heaviest elements (silicon, magnesium, metals) are in solid form at temperatures higher than 1000 K: they form the basis of refractory materials (silicates, oxides, etc.). The heavy elements are formed in the most massive stars, in the latest phases of stellar nucleosynthesis and their fusion requires an enormous energy, which explains their low cosmic abundance.

Within a distance of about 2 AU from the Sun, the temperature is higher than about 200 K, i.e. sufficient for all the small molecules associated with hydrogen (H_2O, CH_4, NH_3 ...) to be in gaseous form. They cannot be incorporated into the solid embryos that will form planetesimals. Only refractory elements can be used for these embryos; this explains why the planets formed at this distance — the telluric planets — are relatively small but with a high density. The atmospheres surrounding them were acquired in a later stage, either by outgassing or by meteoritic/micrometeoritic bombardment; they represent only a negligible part of their total mass.

In contrast, at larger distances from the Sun, the temperature is low enough for the small molecules to be in solid form. The amount of solid matter is thus sufficient to form big nuclei, which may reach about ten terrestrial masses. Theoretical models predict that beyond a critical mass of 10 to 15 terrestrial masses, the gravity field is sufficient to capture the surrounding material — mostly composed of hydrogen and helium. The collapse of this subnebula on the icy core leads to the formation of planets which are very big but with a low density: the giant planets.

Where is the crossover between these two kinds of planets? This is the frontier beyond which small molecules are in the form of ice. Among these molecules, water has a predominant role. First, this molecule is very abundant, simply because of the high cosmic abundances of the two atoms which make it, H and O. Second, as the heliocentric distance increases and the temperature decreases, water is the first molecule to condense, much before the other molecules. Water condensation actually marks the crossover between the telluric planets and the giant planets: it is called the "snowline".

What is the location of the snowline? Today, it is found around 2 AU, corresponding to a temperature of 180 K, i.e. the temperature of water condensation in this environment. At the time of planet formation, the temperature of the protosolar disk was higher. The snowline was probably located between 3 and 4 AU. This explains why Jupiter, the closest giant

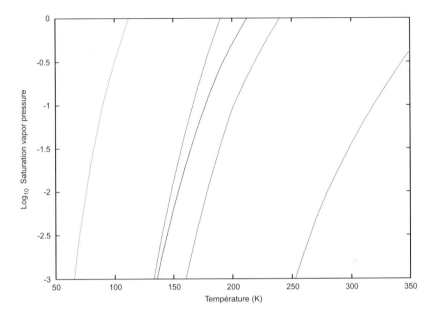

Figure 2.8. Saturation curves of various ices. From left to right: H_2O, NH_3, H_2S, CO_2, CH_4. It can be seen that water condenses at temperatures much higher than those of the other molecules. After T. Encrenaz, *Ann. Rev. Astron. Astrophys, 2008.*

planet to the Sun, is also the most massive one: as it formed just beyond the snowline, it benefited from a large amount of icy particles to build its core.

2.6 Between the two families of planets: asteroids

How can we explain the void between the orbits of Mars, the most remote terrestrial planet, and Jupiter, the closest giant one? Already in the XVIIIth century, two astronomers, Johann Tietz (Titius) in 1766 and Johann Bode in 1772, suggested that the planetary semi-major axes followed a geo-metrical law, according to the formula

$$r = 0.4 + 0.3 \times 2^n$$

n being equal to infinity for Mercury, 0 for Venus, 1 for the Earth, 2 for Mars, 4 for Jupiter, and 5 for Saturn. They had thus noticed that there was no planet corresponding to $n = 3$. After the discoveries of Uranus, Neptune and the transneptunian objects, and with the development of numerical simulation models, the Titius–Bode law has been invalidated. Still, the hole between Mars and Jupiter exists. What can be the reason for it?

The answer to this question came as early as the beginning of the XIXth century, with the discovery of the first asteroids. After Ceres, Pallas, Juno and Vesta, hundreds of objects of this family have been found at heliocentric distances close to 2.7 AU: these rocky bodies, with diameters of a few hundred kilometers or less (the biggest asteroid, Ceres, has a diameter close to 1000 km), form the Main Asteroid Belt. Other populations of small bodies have been also discovered, some closer to the Earth, others at the orbit of Jupiter and beyond. Over 20,000 objects have been discovered today.

What is the origin of the Main Asteroid Belt? Numerical simulations suggest that these asteroids are the remnants of rocky planetesimals which could not accrete into a planet because of the strong gravity field of Jupiter, the most massive giant planet.

Formed beyond Jupiter's orbit, comets are the remnants of icy planetesimals leftover in the process of planetary formation. Their mean size does not exceed about ten kilometers. Located in very elliptical orbits, they occasionally approach the Sun and the Earth. Then, their icy surface is heated by the solar radiation field and the ice sublimes, dragging a cloud of dust which forms the coma and the dust tail. The appearance of the comet at night may be very spectacular, as was the case for comet Hale-Bopp in 1997. Such unexpected apparitions have been the cause of a lot

Figure 2.9. Comet Hale-Bopp. This very bright and large new comet appeared in 1997 (photograph by M. de Muizon).

Chapter 2. The Birth of Planets

of terror and superstition in the old days, until the astronomer Edmund Halley demonstrated the physical nature of comets by successfully predicting the appearance of a periodic comet (later named Halley) in 1859.

Comets are believed to originate from two distinct reservoirs, which can be identified from their orbital properties. After the formation of Uranus and Neptune, under the effect of gravitational perturbations due to Jupiter and Saturn, planetesimals leftover in the process were expelled outside the solar system in a wide shell called the Oort cloud. Occasionally, an Oort cloud comet can be injected again towards the inner solar system. Then, as a result of new planetary perturbations, it can stabilize on a medium or short period orbit; this is the case of Halley's comet, whose period is 76 years. The second reservoir is the Kuiper Belt, located beyond Neptune's orbit (see below, Chapter 2, Section 7). To be complete, it must be mentioned that a third family has been proposed recently. Its origin would be the outer asteroid belt, where some objects have been found to show a cometary activity. These objects might be at the origin of water on Earth (Chapter 3, Section 5).

2.7 Pluto and the transneptunian objects

Since the middle of the XXth century, it has been claimed that some objects might exist beyond Neptune's orbit. The reason proposed by Kenneth Edgeworth, then by Gerard Kuiper, was the sudden drop observed in the matter density distribution of the protosolar disk if no planet was present outside Neptune. Following unsuccessful attempts to find the Xth planet, the two astronomers suggested the existence of a population of small bodies, located in a torus — later called the Kuiper Belt — in the vicinity of the ecliptic plane, too small to be detectable; they called them "transneptunian objects". They also predicted that the Kuiper Belt had to be the reservoir of a specific class of comets, those with short periods and low eccentricity.

These predictions were confirmed in 1992 when two American astronomers, David Jewitt and Jane Luu, discovered the first object, after a long systematic monitoring effort focused on the ecliptic plane. Stellar fields were systematically recorded by visible cameras, then compared with other images recorded later, in order to identify possible moving objects; this is the method used for discovering comets and satellites. Following this first discovery, many detections of TNOs have been discovered; over 1300 TNOs are known today.

It soon appeared that transneptunian objects, depending on their orbital properties, fall in different classes. Most of them, called "classical" objects, have quasi-circular orbits, with a low eccentricity and an inclination lower than 30°; most of them are located between 42 and 47 AU. The second category — the "resonant" or "Plutinos" objects — includes about 12% of the population and shows an interesting property: their revolution period is exactly the one of Pluto, corresponding to a semi-major axis of 39.5 AU. They are all in 3:2 resonance with Neptune, which means that they make exactly two revolutions around the Sun while Neptune makes exactly 3. This orbital configuration is especially stable, which explains the large number of resonant objects; they also have higher inclinations and higher eccentricities than the classical objects. The existence of this family demonstrates the fact that Pluto is indeed a transneptunian object; it is simply among the biggest ones, if not the biggest one discovered so far, and this explains its early discovery.

Finally the last category includes the most distant objects and each discovery is a new surprise. The so-called "dispersed" objects have very elliptical orbits, in strong interaction with Neptune, while the "detached" objects, also on very elliptical orbits, have no interaction with Neptune. Some objects have perihelions beyond 1000 AU... Other discoveries are in front of us, and the frontiers of the solar system keep moving further and further away.

3

Exploring Planet Earth

Several factors play a decisive role in defining the nature of a planet. Some are linked to its orbit, others to its physical properties.

The orbit of a planet determines its distance to the Sun (or its host star), hence the quantity of stellar flux received by the planet. In the case of the Sun, it peaks in the visible range. Part of this flux is absorbed and converted into thermal heat which dominates in the infrared range; thus, the more distant from their star, the cooler the planets.

The solar (or stellar) presence is also found in the form of a solar (or stellar) wind. This flux of energetic particles, mostly composed of electrons and protons, interacts with the planet. The magnetosphere induced by this interaction is especially complex if the planet has a magnetic field and an atmosphere; this is the case of the Earth and the giant planets.

The inclination of the planet's rotation axis, also called obliquity, is another factor which determines seasonal and climatic effects; those are especially noticeable in the case of the Earth and Mars, which have an obliquity close to $24°$. They are responsible for an intense atmospheric circulation and the presence of seasonal polar caps.

In the solar system, planets have low eccentricities. This is not the case, however, for planets around other stars. A high eccentricity is likely to strongly influence the planet's climate by inducing drastic temperature changes at its surface.

Among the physical parameters, the most important one is the mass of the planet, which determines its gravity field, and thus its ability to retain a stable neutral atmosphere. The mass also determines the quantity of internal energy coming from the dissociation of the radiogenic elements included in the nucleus. This energy is responsible for volcanic or tectonic activity at the planet's surface.

The density of the planet depends on its chemical composition. The albedo, i.e. the fraction of reflected solar/stellar light) depends on the surface composition (and the cloud structure).

Finally, the rotation period has an effect on the existence (or not) of a magnetic field. If the planet is a fast rotator, an internal field may be generated in the fluid interior by the dynamo effect. Still there are some exceptions to this rule. Both Mercury and Venus are slow rotators. Yet, Venus has no magnetic field while Mercury has an intrinsic field; its origin is not fully understood yet.

In the following sections, we are going to describe the different components of the Earth system: its internal structure, its atmosphere and hydrosphere, and finally its biosphere which represent a unique case in the Universe, as our planet is presently the only body where life is known to exist.

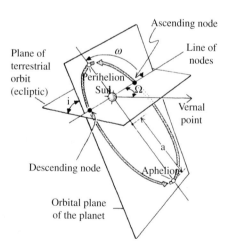

Figure 3.1.   Diagram of a planetary orbit (after T. Encrenaz *et al.*, The Solar System, EDP-Sciences/CNRS-Editions, 2003).

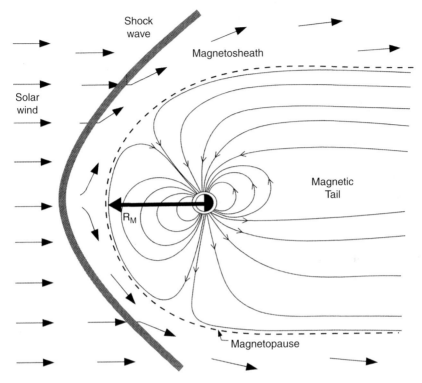

Figure 3.2. Example of a planetary magnetosphere. According to T. Encrenaz *et al.*, The Solar System, ibid.

3.1 Rocks and metals: a differentiated internal structure

Let's start with the Earth' basic parameters: its size and radius. While the terrestrial radius (close to 6400 km) has been known since antiquity, thanks to the work of Eratosthenes, the mass of our planet was not known until the work by Newton at the end of the XVIIth century. Other measurements were achieved, a few decades later, by Pierre Bouguer who studied the shape of the Earth from its gravitational field. The result is a density of 5.5 g/cm³, the highest value found in a solar system body. Let us note that this density is obtained as a direct measurement of the mass/volume ratio: taking into account the compression factor which takes place at deep internal level, this density value is higher than the ones of the material the planet is made of.

Our knowledge of the terrestrial internal structure has decisively progressed during the XXth century, first with the discovery of seismic waves, then with the discovery of continental drift and plate tectonics.

Seismic waves provide us with information about the internal structure of the terrestrial globe. They can be of two types: (1) the internal waves, propagating in the planet's interior with speeds which depend on the nature of the medium they interfere with, and (2) the surface waves, propagating at constant speed. Internal waves can be either pressure waves, propagating in a longitudinal way, or shearing waves, propagating in a transversal way. The latter allow us to determine elastic properties of the materials (rigidity, incompressibility) and do not propagate into fluid media.

Since the beginning of the XXth century, physicists have understood the use of seismic waves emitted during earthquakes for probing the Earth' interior, and an array of seismometers has been set up at a global scale, covering the whole range of frequency emissions from the milli-Hertz to a few tens of Hertz. In 1935, the German geophysicist Charles Richter proposed calibrating the strength of earthquakes following a scale which now bears his name.

Following all information collected by seismometers around the terrestrial globe on distant earthquakes, it appears that the interior of the Earth can be divided into three big regions: the mantle, which surrounds the external liquid nucleus, which itself surrounds the internal solid nucleus (the seed). The mantle and the external nucleus are separated by the so-called Gutemberg discontinuity, at a depth of 2900 km; the external and internal nuclei are separated by the so-called Lehmann discontinuity, at a depth of 5000 km. The study of closer earthquakes has given information about internal regions closer to the surface. These studies have shown, at a depth of about a 100 km, the so-called Mohorovicic discontinuity, between the crust (where waves propagate more slowly) and the mantle. What is the origin of these discontinuities? It is connected to the chemical and mineralogic nature of the elements present in the Earth mantle. As with the other telluric planets, the interior of the Earth is mostly composed of silicates and minerals. This composition is a result of the thermochemical equilibrium, which determined the relative composition of minerals as the protosolar disk started to cool down after its accretion phase (see above, Chapter 2, Section 5).

Silicates can take multiple forms, combining silica and oxygen atoms with others (aluminum, potassium, calcium, iron, etc.). They can also combine with water to produce hydrated silicates. These combinations lead to a great variety of minerals. Besides the silicates, other minerals are also abundant: oxides (as a result of the high relative abundance of oxygen in the Universe), in particular the iron oxide Fe_2O_3. One should also mention calcium carbonate $CaCO_3$, very abundant in the ocean floor.

In order to understand the internal structure of the Earth, we need to complete mineralogy with a geological classification of rocks. Three big families can be identified, depending on the physical processes (temperature, pressure) that they have encountered. The first group is the igneous (or magmatic) rocks, coming from magma crystallization after the partial fusion of mantle rocks. The second group is the sedimentary rocks, formed at the surface of the Earth by superficial erosion or liquid water precipitation. In the third group, metamorphic rocks were formed within the Earth from sedimentary rocks at high temperature and pressure.

Our knowledge of the nature of internal rocks is based on several types of observations: deformation of folded rocks, deep sounding of the terrestrial crust (down to 15 km) or the oceanic crust (down to 3 km), and lava studies, down to depths of a few hundred kilometers. In order to study deeper regions, we can use meteorites, some of which can come from differentiated asteroids, with a composition comparable to the one of the terrestrial interior. Laboratory experiments also allow us to simulate the behavior of minerals at high temperatures and pressures (5000 K, 10^2 GPa, i.e. 1 Mbar). With all these inputs, we can try to describe the main features of the Earth's internal structure.

The composition of the crust is basaltic (i.e. volcanic) in the oceans, and granitic (i.e. metamorphic) on the continents. The thickness of the oceanic crust ranges between 5 and 8 km, while the continental crust is thicker (about 30 km). The crust differs from the mantle below it by its chemical composition. With about three quarters of the total mass of the Earth, the mantle is essentially made of two types of silicates, olivine ($(Mg,Fe)_2SiO_4$) and pyroxene (the simplest one being $NaMg(SiO_3)_2$).

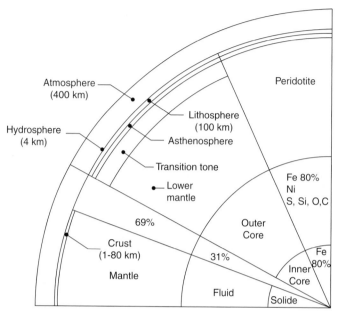

Figure 3.3. Diagram of the internal structure of the Earth (© C. Sotin *et al.* Planetology, Wiley, 2009).

In the upper mantle, at a depth of about 100–200 km, the transition between the lithosphere and the asthenosphere corresponds to the fusion temperature of these compounds.

At a depth of 2900 km, the Gutemberg discontinuity separates the silicated medium, weakly conductive, from the metallic medium, very dense and conductive, mostly made of a fluid iron–nickel alloy. The terrestrial magnetic field is generated by the dynamo effect in the outer part of the nucleus, subjected to the planet's rotation and agitated with convective motions. At even deeper levels, the solid seed has the same chemical composition as the liquid outer nucleus but under the pressure conditions of the medium (365 GPa), the temperature (5500°C) is lower than the fusion temperature.

Can the differentiated internal structure of the Earth be transposed to other telluric planets? Regarding the global chemical composition, the answer is yes, as the initial conditions were similar for the four planets. However, the variations in heliocentric distances have led to significant differences in their internal structures. Mercury is globally denser than the

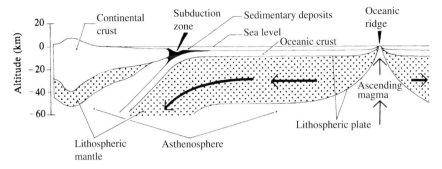

Figure 3.4. Diagram of plate tectonics, showing a section of the lithosphere (after BW Jones, The solar system, Pergamon Press, 1984, © T. Encrenaz *et al.*, The Solar System, ibid.).

Earth (its density is higher when the compression factor is taken into account). The mass of the planet is also an important factor, as it influences the pressure and temperature conditions at its center. The nucleus of Mars is known to have a composition different from the one of the terrestrial nucleus; the origin of this difference is still poorly understood.

3.2 A terrestrial singularity: plate tectonics

Until the end of the XIXth century, it was believed that continents were fixed over the ocean. Still, two types of observations did not support this idea: first, the strange similarity between the coast lines on both parts of the South Atlantic and the Indian Ocean; second, the identity of geological formations shown on continents which are now very far apart. At the beginning of the XXth century, the geophysicist Eduard Suess suggests that all continents were first part of a single one, the Gwondana, and that the missing parts have fallen into the ocean. Alfred Wegener, on the same basis, suggested a modified scenario: the continents, whose crust is lighter than the oceanic crust, could not be immerged, but must have drifted to their present position. In addition to the argument of the coastal similarity, Wegener pointed out that some fossils of rare species were found on continents which are now separated; he also mentioned the tropical distribution of some glacial features of the secondary era.

The visionary intuition of Alfred Wegener was based on clear observations, but the physical explanation of the motion mechanism was still

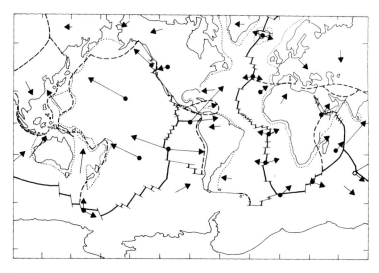

Figure 3.5. This map shows the motions of the main plates relative to fixed reference hotspots. According to C. Allègre, The scum of the Earth, 1983; © T. Encrenaz *et al.*, The Solar System, ibid.).

missing, i.e. the identification of the energy source required to generate the drift of continents. This is why Wegener's theory was not generally accepted until the second half of the XXth century. In the 1960s, the geologist Harry Hess suggested that the submarine oceanic ridges, discovered by submarines during the Second World War, are the place where the mantle material is rising to form the oceanic lithosphere, while an equivalent quantity of matter is returning into the mantle at the level of oceanic trenches, below the continental crust. The mechanism is globally generated by convection at the level of the upper mantle. The lithosphere is made of five big plates, separated by unstable regions where seismic and volcanic activity takes place.

Since the work of Wegener and Hess, many observations have helped to confirm the plate tectonics theory and to understand its mechanisms. These measurements deal with the remnant magnetic field fossilized in the rocks, which give information of its intensity and orientation at the time of their formation. As an example, measurements performed in the Deccan Traps have shown evidence for the mobility of the Indian ocean, presently drifting northward. The magnetic inversions found in the oceanic crust can be explained by the uplifting of magmatic matter along

Chapter 3. Exploring Planet Earth

the ridges, and allows us to trace the history of the oceanic crust, whose age is no more than two hundred million years. Finally, the plate tectonics model gives a simple explanation to the origin of the volcano chain of Hawaii: the volcanoes were formed sequentially as their lithospheric plate moved, in the north-west direction, over a fixed hot spot, a sort of vertical chimney originating inside the mantle; its episodic activity triggers the formation of a volcano.

The plate tectonics mechanism finds its origin in the convective motion of the upper mantle. Can such a process occur on other solar system objects? Signs of tectonic activity are found in many surfaces of planets and satellites, but no mechanism equivalent to terrestrial plate tectonics has been found. The reason is that such a mechanism requires an important energy source, mostly originating from the dissociation of radiogenic species. The Earth, being the largest rocky body in the solar system, is probably the only one having the critical mass. It seems also that the plate tectonics mechanism is made easier by the presence of liquid water, as it favors convection in the upper mantle.

3.3   Formation of terrestrial relief: the isostasy principle

What is the origin of the terrestrial mountains? The horizontal motions of the lithospheric plates are responsible for vertical motions of smaller amplitude, especially when two continental plates come in contact; this process can induce a local folding of the continental crust. This is how, in Europe, the Hercynian folding occurred in the Palaeozoic era, and then the Alpine folding in the Tertiary. In the same way, the Himalayas emerged after the subduction of India under the Eurasian plate. Another source of relief is volcanism, also a consequence of plate collisions, in particular at the border between oceanic and continental plates: this is the case of the Cordillera of the Andes. As mentioned above, another type of volcanism is associated with Hawaiian volcanoes that originate from a deep hot spot located in the mantle.

However, there is a limit to the altitude that terrestrial mountains can exceed. Indeed, the continental crust is floating on the lithospheric mantle, and the weight that can be supported by the mantle is constrained by

Archimedes' principle. In case of an overload, the crust sinks: this is the isostasy principle, which determines the maximum altitude of the terrestrial mountains. This altitude depends on the gravity and the relative densities of the oceanic and continental crusts (2.7 and 2.9 g/cm³ respectively) that float on the upper mantle (3.2 g/cm³). Using these values, we can estimate that the maximum contrast between the highest summits and the deepest trenches is about 6000 m. The observed value is higher, since the highest summits of Himalaya are above 6000 m while the deepest trenches, the Mariana trench, is at a depth of 11,000 m. This large difference is partly explained by the presence of the volcanic islands of the Pacific ocean, and also by the fact that the youngest mountains, in the Alps in particular, have not yet reached isostatic equilibrium.

In the same way, on each planet or satellite, there is a maximum altitude that cannot be exceeded by the reliefs. The more massive is the planet, the more active volcanism may be; in contrast, in the case of a low-mass planet, with a low gravity field, the ratio between the maximum altitude and the planetary radius is higher. This is why Mars, which is ten times less massive than the Earth, has very large volcanoes with an altitude exceeding 30 km, if compared with the deepest layers; these are the highest mountains known in the solar system.

3.4 Our natural environment: the terrestrial atmosphere

With its seasonal variations, the terrestrial atmosphere, interfaced between the surface and the interior on one side, and on the interplanetary medium on the other side, has an overwhelming influence on our daily life. The main parameters are the temperature and the pressure; they determine the winds and the transition changes between the different phases of water. The consequences of these changes are, in particular, the glacier formation, the cloud structure and the fluvial network.

What do we know about our atmosphere? From the Greek "atmos" (vapor) and "sphaira" (balloon), this word describes the gaseous envelope surrounding a celestial body. The terrestrial atmosphere has been studies for several millennia. The Chinese, then the Greeks, among them Thales and Aristotle, have attempted meteorological observations. It was only at the beginning of the XVIIth century, with Galileo, that the first (still imprecise) measurements of the air density were achieved. Torricelli, in 1643, realized the first measurements of the atmospheric pressure. Then, new instruments were conceived: the hygrometer for measuring water content, the anemometer for measuring wind speed, the thermometer for temperature measurements. At the end of the XVIIIth century, the first measurements of the atmospheric composition, with about 78% nitrogen and 21% oxygen, were obtained; minor species like water vapor and carbon dioxide, more easily detectable, had been previously identified. At the same time, several chemists (Jacques Charles, Amedeo Avogadro, Louis Gay-Lussac, John Dalton) discovered the perfect gas law, which determines the relationship between the pressure of a gas, its temperature and its volume. The first measurements at high altitude, in particular on Mont Blanc, allowed scientists to determine the variations of the temperature and pressure as a function of altitude. At the beginning of the XXth century, radio probes allowed them to reach higher altitudes; then, in the second half of the XXth century, the advent of the space era allowed us to probe the atmosphere up to its interface with the magnetosphere and the interplanetary medium.

In parallel, astronomers have discovered that, with the exception of Mercury, all solar system planets are surrounded with an atmosphere. Carbon dioxide is identified as the main atmospheric component of Mars and Venus. Methane and ammonia were detected on Jupiter, as well as methane on Saturn, as early as 1932. These constituents are very minor in the giant planets, but their spectrum is easier to identify that the one of hydrogen, the main atmospheric component, which was discovered only in 1960. After this date, planetary space exploration have provided us with a more complete description of solar system planetary atmospheres.

Let us briefly describe the main characteristics of the terrestrial atmosphere, in comparison with the others. Mercury, as mentioned above, has no stable atmosphere, as its gravitational field is too small for the planet to retain gaseous molecules, even the heaviest ones. The terrestrial atmosphere appears as an intermediate case between the ones of Venus and Mars, where the surface pressures and temperatures reach extreme values: at the surface of Venus, the temperature is as high as 730 K

(457°C) and its pressure is 90 bars (9 10^6 Pa); in contrast, the mean surface pressure on Mars is about 0.006 bar (600 Pa) and its mean surface temperature is about 210 K (–63°C); both parameters strongly vary as a function of season. Another surprise comes from the atmospheric compositions: the atmospheres of both Venus and Mars are mostly made of carbon dioxide, with a few percent of molecular nitrogen, and traces of other constituents (CO, H$_2$O ...). It is most likely that the primitive atmosphere of the Earth had a comparable composition, with, in addition, large amounts of water. But a decisive difference took place: on Earth, the temperature was such that water could remain in the liquid form, and carbon dioxide was trapped under the oceans in the form of calcium carbonate. Molecular oxygen appeared later as a consequence of the origin of life, which led to the present atmospheric composition, dominated by nitrogen

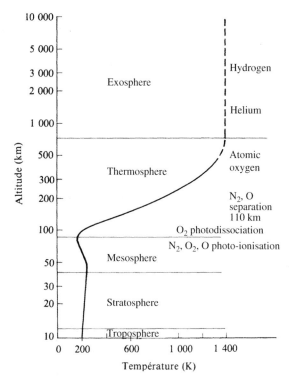

Figure 3.6. Diagram of the thermal profile of the atmosphere (© T. Encrenaz *et al.*, The Solar System, ibid.).

and oxygen. What are the reasons for the diverging fates of Venus and Mars? Answering this question is one of the major challenges of today's planetology (see Chapter 5, Section 5).

How can we describe the terrestrial atmosphere? Several laws determine the evolution of its pressure and temperature as a function of altitude. The first one is the "hydrostatic law" which expresses the balance between the pressure (i.e. the weight of gas above the given level) and the gravity (associated with the Earth mass). As altitude increases, the pressure decreases following a law that is, to first order, exponential. On Earth, the pressure decreases by a factor e (e ~ 2.7) when the altitude increases by 8 km called one scale height. On Mars, the scale height is typically 10 km; at the surface of Venus, the scale height is 14 km. The scale height is proportional to the inverse of the gravity and the molecular mass; it is also proportional to the atmospheric temperature as expressed in Kelvins.

In the largest part of the atmosphere, the various atmospheric constituents are uniformly mixed, whatever their molecular masses are; this is the homosphere which, on Earth, extends up to an altitude of 90 km. At this level, the homopause separates the homosphere from the heterosphere that extends above it; the pressure at this level is a few microbars (or tens of Pa). In the heterosphere, the various species separate, each following its own scale height, determined by its molecular mass.

The vertical temperature distribution in the Earth's atmosphere is complex, and driven by different mechanisms. As in the case of Mars and Venus, the temperature above the surface starts to decrease as the altitude increases; we are in the troposphere. The surface is heated by the solar visible radiation and the heat is transferred to the atmosphere through convection. The temperature decreases by about 6 K/km, corresponding to the adiabatic gradient. This regime prevails up to the tropopause, at an altitude of about 12 km; at this level (the tropopause), the temperature is at a minimum with a value of 217 K (−56°C) then stays isothermal above this level. Above this isothermal region, at a level of about 25 km, the temperature increases again with altitude, due to the absorption of the solar UV radiation by molecular oxygen which dissociates to form the ozone layer: this is the stratosphere. Above it, in the thermosphere, the temperature decreases again as altitude increases, down to a second minimum (170 K, or about –100°C, at about 90 km). This region is overlaid by the thermosphere where the temperature increases again sharply, due to the interaction of the atmospheric atoms with energetic particles coming from the solar wind; this is the region where auroras form.

Why does the temperature vary with altitude? The main reason is the absorption of solar radiation by atmospheric constituents at different atmospheric layers. The UV radiation, very energetic, ionizes the atoms and dissociates radicals of the upper atmosphere, while the near-infrared solar radiation excites the neutral molecules at lower altitudes. This is how the ozone layer is formed in the Earth atmosphere, by absorption of the solar UV flux. This phenomenon has no equivalent on Mars and Venus, because oxygen in these planets is only present as a trace constituent. In contrast, the photodissociation process is found in giant planets, where methane is dissociated in the stratosphere to form radicals that, in turn, recombine in several hydrocarbons.

In the troposphere, the heating of the atmosphere comes from the infrared radiation of the surface itself, warmed up by the solar radiation in the visible range. The infrared atmospheric radiation, in turn, heats the surface, raises its temperature and accelerates the mechanism: it is the

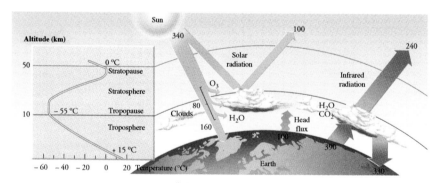

Figure 3.7. Diagram of the greenhouse effect: The visible solar radiation penetrates through the atmosphere and heats the surface, which emits infrared radiation toward the lower atmosphere. Some atmospheric gases (CO_2, H_2O, CH_4) absorb this radiation, which increases the heating of the lower atmosphere and the surface, and the mechanism tends to accelerate. (© T. Encrenaz, Planetary atmospheres, Belin, 2000).

Chapter 3. Exploring Planet Earth

greenhouse effect. Its effectiveness depends on the nature of the gases in the atmosphere. Water vapor, carbon dioxide and methane have a particularly rich infrared spectrum that allows them to absorb infrared radiation: they are very effective greenhouse gases. In contrast, molecular nitrogen and oxygen are ineffective. On Venus, whose primitive atmosphere had to be rich in carbon dioxide and water vapor, the greenhouse effect has been very active, hence the very high surface temperature observed today (see Section 4.3.2). On Mars, whose atmosphere is extremely tenuous, the greenhouse effect is very limited. On Earth, the greenhouse effect is

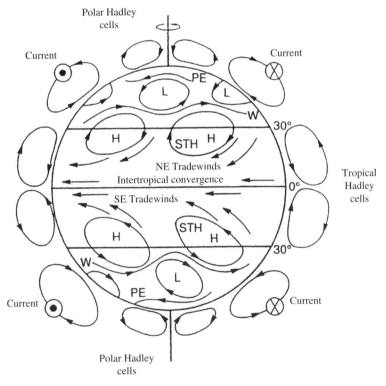

Figure 3.8. The terrestrial atmospheric circulation, known as Hadley circulation, consists in several convective cells, symmetrical with respect to the equator. At the equator, where the Sun is at zenith, the hot air rises and forms high altitude clouds. The air is transported to the tropics where it descends into dry areas. Other convective cells are formed at higher latitudes, due to the Coriolis forces associated with the rotation of the planet. (© Encrenaz *et al.*, The Solar System, ibid.)

modest because the water is in liquid form, and most of the carbon dioxide is trapped in the oceans.

The temperature difference between the surface and the troposphere causes convective motion, ensuring the transport of energy to higher altitudes. Near the equator where the surface temperature is maximum, hot gaseous masses rise, cool at higher elevations and then fall down at higher latitudes, in a relatively stable cycle called the "Hadley" cycle. We find the same cycle on the other planets. If the Earth itself had no rotation, a single convective cell would be installed on each hemisphere and air masses would move from the equator up to the polar regions; this is what we observe on Venus, whose rotation speed is almost zero. In the case of the Earth, the rotation of the planet generates a Coriolis force, which causes fragmentation of each hemispherical cell into three components. The upward movement is accompanied by the formation of clouds of water vapor and precipitation, as observed in the equatorial regions. The dry air masses descend in the tropics where desert regions dominate.

3.5 The water cycle: another specificity of Earth

On Earth, water exists in three phases, liquid, solid and vapor; this feature of our planet is so far unique in the Universe. Indeed, while the H_2O molecule is ubiquitous in stars and the interstellar medium, from the solar system to distant galaxies, we always find it in the form of gas or ice. However, it seems that liquid water was present on Mars in the past history of the planet (see Section 4.4.2), and liquid water could also exist in the interiors of some of the outer solar system satellites. But the coexistence in the same medium of the three phases of water appears as a singularity of our planet.

The presence of liquid water on Earth was of fundamental importance for the evolution of our planet: it helped reduce the greenhouse effect by removing at the bottom of the ocean almost all the carbon dioxide, and thereby regulating the air temperature during its history. Liquid water has undoubtedly played a key role in the emergence of life and it is now essential to its preservation. On the geophysical front, the presence of oceans has multiple fundamental consequences. The three phases of water are exchanged in a cycle that encompasses the entire planet. Ocean surface evaporates and water vapor is transported by the wind until it condenses into rain or ice over the continents to form glaciers and rivers which drive water back to the seas. The seawater is salty, as a result of the contribution of oceanic ridges and continental waters that have been enriched in salt by their contact with soil and rocks. By evaporating, the seawater discharges its salts and precipitates as freshwater. The Earth's oceans are subject to a surface circulation driven by wind and coupled with atmospheric circulation (especially in the southern hemisphere, where there is less land), but also to a deep circulation, which depends on the temperature and salinity of the medium.

The water cycle has a direct impact on the terrain and topography of the ground. On the one hand, the ice accumulates in the form of polar ice caps: at the north pole, there is the floe whose extension is seasonal while, at the south pole, it permanently covers the Antarctic continent. The snowfall also feeds the glaciers digging the surface of the Earth into broad valleys. On the other hand, liquid precipitations feed the river systems that also mark the surface. In terrestrial deserts, in the Sahara or in Yemen, traces of currently dry networks are found, showing evidence for different climatic conditions in the past. Many traces of glacial erosion down to the mid-latitudes are relics of the period corresponding to the quaternary glaciation. Remarkably, traces of fluvial and glacial erosion have been identified on the surface of Mars, which reinforces the hypothesis that liquid water flowed on the surface of Mars (see Section 4.4.2). It seems that in the past, the three phases of water coexisted on planet Mars, but we do not know with certainty at which time, nor how long it lasted, as we do not know either why it ended (see Section 4.4.3).

If the cycle of water in its three phases observed on Earth today is, as far as we know, unique in the solar system, Saturn's satellite Titan exhibits a similar situation, in which methane CH_4 plays the role of water. The temperature is very low (93 K at the surface, or $-180°C$), but the atmosphere of Titan shows some similarities with the Earth: it is composed mainly of molecular nitrogen and its surface pressure is 1.5 bar; the methane is present at the level of a few percent. The minimum atmospheric temperature is about 70 K, which is below the condensation temperature of methane. Moreover, we know, thanks to images of the Cassini–Huygens probe, that traces of fluvial erosion are very numerous on the surface of Titan.

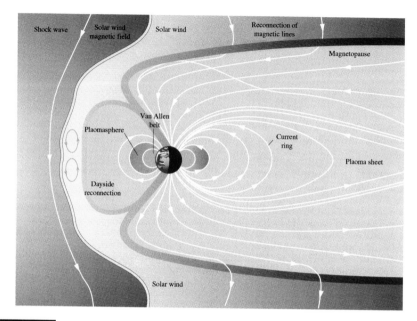

Figure 3.9. Diagram of the Earth's magnetosphere. Its high activity is due to the magnetic field amplitude and the proximity of the planet to the Sun. (© T. Encrenaz, Planetary Atmospheres, Belin, 2000).

Boulders, most likely composed of water ice, appear heavily eroded; the cause seems to be a mixture of hydrocarbons, probably based on methane, which has flown in these valleys. In addition, the radar measurements from the Cassini spacecraft have discovered that the satellite is partially covered with lakes of methane at high latitudes. It is very likely that there is a cycle on Titan during which methane escapes from the inside then condenses in the upper troposphere to precipitate at higher latitudes, according to a seasonal cycle somewhat analogous to the one we know on Earth. Regarding the stones of water ice, they would be the equivalent of our silicate rocks.

3.6 Between the Earth and space, the magnetosphere

In the upper atmosphere, the auroral phenomena arise from the interaction between the solar wind and the magnetic field, generated by a dynamo effect in the liquid core composed of iron and nickel. This field can be

Chapter 3. Exploring Planet Earth

represented by a dipole slightly inclined (14°) relative to the axis of rotation of the globe. The magnetosphere results from the interaction of the magnetic field with ions and protons in the solar wind ejected from the solar corona and accelerated to supersonic speeds.

The terrestrial magnetic field can be considered an obstacle, which deflects the solar wind and forms a magnetic cavity, the magnetosphere. Located at ten Earth radii in the direction of the Sun, the magnetopause marks the boundary between the solar wind and the magnetosphere. Upstream of the Sun–Earth axis appears another surface of discontinuity, the shock wave, where the solar wind is compressed, decelerated and deflected to either side of the magnetosphere. The magnetosphere has an overall symmetry with respect to the Sun–Earth axis but is strongly asymmetric with respect to the center of the globe: a magnetotail develops in the direction opposite the Sun up to several tens of Earth radii. The magnetosphere plays an important role as an interface between the upper atmosphere and the interplanetary medium: by shielding the solar wind, it protects the atmosphere by making escape mechanisms less effective.

In the ionosphere, the atoms are ionized not only by the ultraviolet and X radiation from the Sun, but also by energetic particles from the solar wind. The particles, accelerated along the magnetic field lines, rush to the polar regions in two symmetrical cones. In these regions, auroras are generated by the radiation emitted in the visible range from atoms and molecules excited by energetic particles. As another effect of the Earth's magnetosphere, the Van Allen belts, located a few Earth radii from the planet, are a network of field lines connecting the northern and southern hemispheres at high latitudes.

Are there other magnetospheres around the solar system? The answer is yes, for some of them. The most similar to ours is the magnetosphere of Jupiter, which has, like the Earth, a magnetic field (also related to its fast rotation) and a dense atmosphere (see Section 5.1.6). The other giant planets also have magnetospheres, each with its own characteristics; Uranus and Neptune, for example, have their magnetic axis strongly inclined with respect to the axis of rotation.

The interaction between the solar wind and the planets may take different forms, depending on whether or not the planet has a stable atmosphere, and the presence or absence of a magnetic field. The Moon is an example of body that lacks both of these two components. We have seen that planet Mercury is devoid of atmosphere but, curiously, it has a magnetic field, which is unexpected in view of its very slow rotation (see Section 4.1). Venus, also in very slow rotation, is — logically — devoid of magnetic field but its atmosphere generates a third type magnetosphere. Another feature is shown by planet Mars, which has no magnetic field today; however, its oldest rocks have kept a record of fossil magnetism, arguing for the existence of a past magnetic field — and thus a dynamo effect — in the early history of the planet. How and why did this field disappear? The question remains open (see Section 4.4.3). It is possible that the disappearance of the Martian magnetosphere led to the escape of the primitive atmosphere, which would explain the very low ground pressure today. Indeed, the study of planetary magnetic fields presents a variety of cases, providing us with many clues about the past and recent history of the solar system.

3.7 A brief history of the Earth's climate

If we try to understand the origin and evolution of the solar system, we must first dig into the archives of the climate of our own planet to decipher its history. What tools do we have? Firstly, the measurements of chemical composition, elemental and isotopic ratios performed on the oldest continental rocks, but also on samples taken from deep ice cores or collected on the ocean floor. Another source of information comes from the periodic reversal of the Earth's magnetic field, fossilized in some igneous and sedimentary rocks that can thus be dated.

What do we know of the history of the Earth? Its birth, contemporary with that of other planets, dates back to 4.56 billion years ago. We have seen (Chapter 2) that the planets of the solar system were formed by accretion of solid particles in a rotating disk. First, embryos were formed and grow as a result of multiple collisions, and later by their own gravitational field. The first few hundred million years were marked by the permanent bombardment of the young planet by celestial bodies of all sizes. The temperature rose, causing volcanic activity. A gradual decrease in temperature then caused the condensation of the oceans; continents appeared about four billion years ago.

At the beginning of its history, the Sun was significantly fainter than today, as we have learnt from the observation of young stars. A temperature lower than today would have resulted in a global glaciation, if the primitive atmosphere had not included a massive amount of carbon dioxide, as in the case of Venus and Mars. The greenhouse effect that has resulted has allowed the Earth to maintain its liquid water. In turn, the presence of liquid water has prevented the runaway greenhouse effect that we observed on Venus. Indeed, carbon dioxide gradually dissolves in the oceans to form calcium carbonate, i.e. limestone. This mechanism has allowed the Earth to maintain a "relatively" constant temperature throughout its history.

The first signs of marine life appeared in the oceans more than 3.5 billion years ago; the mechanisms, endogenous or exogenous, which led to the emergence of life are still a mystery. The land records show two periods of glaciation, one around 2.5 billion years ago, the other around 0.8 billion years ago. When the Earth gradually freezes, its albedo (the ice) increases, which means that the amount of solar energy it absorbs decreases; the phenomenon can therefore accelerate, to lead to a total global glaciation. One can imagine that intense volcanic activity, releasing massive amounts of greenhouse gases such as water vapor and carbon dioxide, allowed the Earth, at least on two occasions, to escape final glaciation.

The emergence of life on the ocean floor leads to a gradual increase in the amount of molecular oxygen in the atmosphere. The first consequence is the oxidation of rocks: iron oxides are found, dated as 1.8 billion years old. Another result that we have already discussed: atmospheric oxygen is dissociated by solar UV radiation to form ozone. This is of fundamental importance for the evolution of life on Earth: by blocking solar UV radiation, which is able to destroy all organic molecules, ozone allowed life to develop on continents. This fundamental transition occured about 540 million years ago.

The gradual increase in solar flux, which compensates for the decrease of the greenhouse effect related to CO_2 sequestration in the ocean, provides some stability conditions. We are entering the second era,

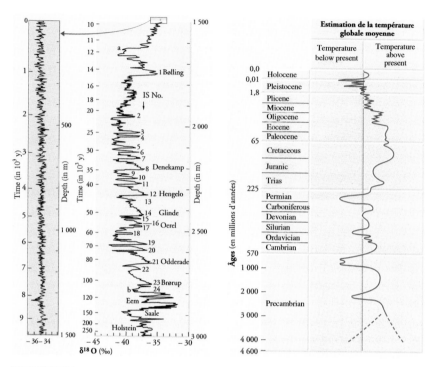

Figure 3.10. Evolution of the temperature in the Earth's history, reconstructed from geological observations (after S. Jousseaume, Climates from yesterday to tomorrow, CNRS-Editions/CEA, 1993).

characterized by a relatively high temperature, which results in the appearance of giant plants, amphibians, insects and reptiles. In the Carboniferous era (from 360 My to 295 My ago), large reservoirs of fossil carbon were formed, coming from the decomposition of plants in swamps. Continents were combined in a single element, Pangaea. Due to melting ice and rising sea levels, the major sedimentary basins were formed. The Cretaceous period marked the end of the Mesozoic era, and the gradual rise in temperature led to the emergence of new plant and animal species, including the dinosaurs and birds. This period also saw the gradual opening of the Atlantic and the Indian Ocean.

Then there came, 65 million years ago, the famous Cretaceous–Tertiary transition, marked by an ecological disaster, the disappearance of dinosaurs and many other species. It is now widely accepted that the fall of a giant meteorite, a dozen kilometers in diameter, in the Chixchulub

Chapter 3. Exploring Planet Earth

region in Mexico, was the cause of this disaster. It is not excluded that intense volcanic activity also helped.

During the Tertiary, continental drift continued: Australia separated from Antarctica and, about 40 million years ago, India became embedded under the Eurasian plate to form the Himalayas. Mammals evolved and diversified (ancestors of horses, mammoths) until the appearance of the first hominids, several million years ago. Then came, about 2 million years ago, a new cooling period: we are entering the Quaternary, marked by a series of successive glaciations. The last one, about ten thousand years ago, expanded from high latitudes to mid latitudes in Europe. The alternation of these successive glaciations was explained by the mathematician Milankovitch as due to changes in the Earth's orbital parameters (eccentricity, obliquity, precession), themselves influenced by gravitational perturbations related to neighboring planets with periods of tens of thousands of years. However, smaller scale variations remain mysterious. Between 8000 and 6000 years ago the average temperature increased again, causing the development of vegetation in the Sahara. Then, 5000 years ago, falling temperatures caused the desertification. More recently, another "Little Ice Age" took place between the fifteenth and nineteenth

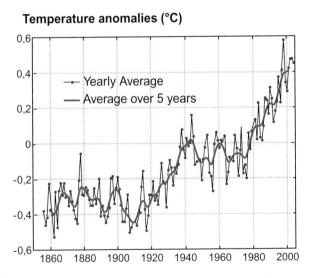

Temperature anomalies (°C)

Figure 3.11. Temperature increase as a function of time during the last 150 years (© Globalwarming, 2007 & IPCC AR4).

centuries, ending the "Medieval Warm Period" that had enabled the colonization of Greenland by the Vikings.

What is striking in the history of Earth's climate, compared to Venus and Mars, is the relative stability of the temperature. It is known that the atmosphere of Venus has heated up considerably during its history, while, in contrast, that of Mars cooled and dissipated into space. What are the factors that are responsible for this stability? The first is undoubtedly the presence of a liquid ocean: the Earth is located at just the right distance from the Sun to keep its surface temperature between 0°C and 100°C. But this factor alone was probably not enough: the balance between the increasing solar flux and the gradually decreasing atmospheric CO_2 probably ensured the proper balance. Perhaps planet Venus, at the very beginning of its history, housed a liquid ocean, but its presence close to the Sun, in this case, caused its evaporation, hence enhancing the greenhouse effect through large amounts of CO_2 and H_2O gas.

3.8 The Earth–Moon couple: a double system

Another key factor for the stability of Earth's climate was probably the presence of the Moon. Celestial mechanics calculations have shown that the presence of our satellite had the effect of stabilizing the obliquity of the rotation axis of the Earth. In the absence of satellite, it would have varied periodically: calculations show that, in the case of Mars, it can reach 60°. If the Earth's rotation axis had experienced such oscillations, induced climatic changes would undoubtedly have been considerable.

The Earth–Moon pair is unique in the solar system. The size of the Moon and its density are different from other satellites and closer to the class of terrestrial planets; in this sense, we can consider the Earth and the Moon as a double system. Observed since antiquity, studied since Galileo, the Moon was the subject of an intensive campaign of space exploration in the 1960s and 1970s, led by the United States (with the Apollo program for the manned exploration of the Moon) and the Soviet Union (with the robotic program for sample collection). The stakes then were mostly political, but science has been able to draw the greatest profit. The result is undoubtedly a major scientific achievement: the dating of lunar samples, returned to Earth for chemical and mineralogical analysis. The *in situ* measurements also allowed the determination of the mineralogical composition of the surface, a measurement of the magnetic field (which was found to be very low), a survey of its internal structure and finally a measurement of the composition of the solar wind; the energetic particles emitted by the Sun were trapped on an aluminum foil placed on the lunar surface and then returned to Earth.

We have seen (Chapter 2) that the scenario of formation of planets in the solar system does not support the formation of satellites around the terrestrial planets, because the accretion phase of the rocky core is not followed by a phase of collapse of the surrounding material in an equatorial disk (as is the case for the giant planets). Then what is the origin of the Moon? In addition to its relatively high mass, the formation scenario must account for its density (3.3 g/cm^3), significantly lower than that of the Earth. Several models have been proposed following the space exploration of the Moon in the 1970s: fission of part of the mantle, with a density comparable to that of the Moon, capture by the Earth of a Moon formed elsewhere independently of the Earth, and accretion in an orbit around the Earth. From the 1980s, the development of simulation models has allowed us to propose a scenario that is now widely accepted by the scientific community. It is now believed that the Earth–Moon system was born from the side impact between the young Earth (aged tens of millions of years) and a smaller planet, the size of Mars. This collision had the effect of causing the fusion of the two nuclei, rich in heavy elements, and injecting into Earth orbit parts of the mantles of the two bodies. Orbital debris, of lower density, were then assembled to form the Moon. This scenario has the advantage of explaining the current density of the Moon but also its initial orbit, highly elliptical and inclined, as can be deduced from its current rate of recession using dynamic models. The Earth–Moon distance increases currently by 4 cm/year due to tidal effects, also responsible for the synchronous rotation of the Moon. In the case of Mars, the same tidal effect is responsible for the gradual recession of the satellite Deimos, and the gradual approach of the other satellite Phobos.

3.9 The Earth, a unique planet...

This short overview of the history of the Earth and its satellite provides us with a first lesson: our planet is unique in the solar system. Its evolution, from its origin to what we know, is the result of a combination of multiple factors, some of which are difficult to quantify.

Initially, the Earth was at a heliocentric distance compatible with the presence of liquid water under the conditions of the current solar flux. But we have seen that the radiation of the young Sun was low, so the presence of a primitive atmosphere rich in carbon dioxide (and possibly methane) was needed to ensure a sufficient greenhouse effect, allowing the Earth to escape complete glaciation. Very early in the history of the planet, a collision occurred with a protoplanet of the size of Mars; as a result, a large satellite was formed, capable of stabilizing the obliquity of the Earth's rotation in the long term.

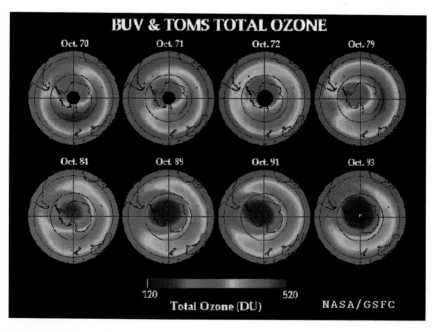

Figure 3.12. The evolution of the ozone hole above the South Pole between 1970 and 1993, measured from satellite observations. Over the past ten years, global actions to limit the rate of CFC have helped to stop the decrease of polar ozone in winter (© NASA/GSFC).

Chapter 3. Exploring Planet Earth

Then life appeared in the oceans, but we still do not know if it came from inside, near hydrothermal vents, or outside, from prebiotic molecules brought by meteoritic and cometary impacts. The release of oxygen resulted in the formation of the ozone layer and made possible the development of life on the continents. On several occasions, periods of total glaciation could have frozen the planet indefinitely as a final snowball, but its internal energy must has generated sufficient volcanic activity to replenish the greenhouse effect. Sixty five million years ago, an ecological disaster probably indirectly favored the emergence and the development of mammals, and, after multiple diversifications, the human species.

What is the future of planet Earth? We are now at a turning point in the history of the Earth's climate. Over the last century, human activity has changed it significantly. Industrial activities are the cause of an increase in the atmospheric CO_2 and therefore the temperature. Another impact of industrial activity on the environment has been the use of freon or chlorofluorocarbons (CFCs), which resulted in the infamous "ozone hole". CFCs, emitted in large quantities at the beginning of the industrial era, have indeed the power of destroying stratospheric ozone molecules because they catalyze photochemical reactions leading to the destruction of O_3. Following the discovery of the "ozone hole" in the early 1980s, international decisions were taken to eliminate the use of CFCs, and it is hoped that the phenomenon is now receding. The problem of global warming is much more serious, because the time constants are very long: even if it were possible today to stabilize CO_2 emissions, the effects would not be felt for a hundred years. Numerical models have been developed to try to predict the expected temperature rise in the coming century. Uncertainties are large, but in all cases, the predictions are very disturbing. The temperature could rise by a few degrees by 2030, and sea levels could rise by a few feet, causing the immersion of many coastal areas. The climatic consequences would be significant: during the Little Ice Age, the average temperature was lower than the current temperature by only 1°.

Since the last couple of decades, humanity has become aware of the seriousness and urgency of the problem caused by global warming. Further action is needed on energy conservation, the search for alternative energy sources, and environmental protection.

4

The Neighbors of the Earth

Mercury, Venus, Earth and Mars: four planets belonging to the same category, the terrestrial planets (also called "rocky" planets), but which have nevertheless extreme differences. What are the similarities and differences between the terrestrial planets, and what are their origins? This is the subject of this chapter.

The common characteristics of terrestrial planets arise naturally from their formation (see Chapter 2). Born relatively close to the Sun from solid material, refractory terrestrial planets have a high density, but are small, due to the limited abundance of this material. Their gravity field was not sufficient to capture the surrounding protosolar gas as, at greater distances from the Sun, the giant planets did. The atmosphere surrounding them is therefore only a tiny fraction of their mass. Even in the case of Venus, which has the densest atmosphere, its mass is below 10,0000th of that of the solid planet. Terrestrial planets acquired their atmosphere partly by outgassing, but probably mostly due to meteoritic or micrometeoritic bombardment by asteroids and comets. Another specificity of the terrestrial planets: they have little or no satellites; existing ones — the Moon around the Earth, Phobos and Deimos around Mars — were not formed at the same time as the planet, but appear to be the result of a collision or a capture. Finally, there is another similarity between the three planets with an atmosphere: the primordial atmospheric composition, based on carbon dioxide, water and nitrogen, was probably similar; in the case of the Earth, the current composition of its atmosphere results from the presence of liquid water and the emergence of life (see Section 3.7).

Now, let us turn to the differences between the terrestrial planets. Mercury is put aside, too small and too close to the Sun to maintain a stable atmosphere. The other three planets have surface conditions which vary between the extreme hell of Venus, with a surface pressure of about 90 bar and a temperature of 457°C, and the deserts of Mars, where the ground pressure is less than one hundredth of bar and where the surface temperature may range from 0°C to −100°C between the equator and the poles. Between these two extremes, the Earth has conditions of temperature and pressure which can keep water in liquid form, and this is what allows us today to talk about it.

4.1 Closest to the Sun, Mercury

Located at less than 0.4 AU from the Sun, Mercury is difficult to observe from Earth, because its angular distance from the Sun is always less than 30°; the planet appears as a thin crescent, visible at dawn or dusk. Space

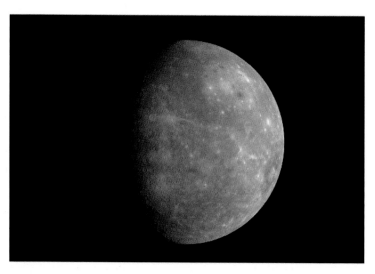

Figure 4.1. Planet Mercury as seen by the Messenger probe (© NASA).

exploration is necessary, and the first step took place in the 1970s with the U.S. mission Mariner 10. More recently, the Messenger mission, launched in August 2004, also by NASA, flew over the planet in January 2008, October 2008, and September 2009, and is now in orbit around the planet.

The rotation period of Mercury (59 days) is in 2:3 resonance with its orbital period (88 days), i.e. the planet rotates three times on itself while making two revolutions around the Sun; this ratio corresponds to a dynamically stable configuration of the Mercury–Sun system. Mercury does not always show the same side to the Sun, as does the Moon with respect to the Earth. Due to the low gravitational field and the high temperature of the dayside, Mercury cannot retain a stable atmosphere. Its surface, heavily cratered, resembles nothing as much as that of the Moon. It reflects the intense meteorite bombardment that took place at the end of the first billion years after the formation of planets (see Section 2.3.3). Apart from these impacts, the surface of Mercury has not changed since the origin of the planet, because the internal energy associated with its mass was insufficient to generate tectonic or volcanic activity.

The lack of atmosphere results in very high temperatures on the dayside (near 700 K, or more than 400°C), but very low ones on the nightside (90 K, about −180°C). In particular, due to the absence of obliquity, polar regions may find themselves permanently in the shade. The temperature would be low enough to keep in the form of ice residues from impacts of cometary origin, rich in water. Such deposits could have been detected by the radar data of the Messenger mission. Bepi Colombo will be able to confirm or refute this result.

Mercury's density is remarkably high. It is almost equal to that of the Earth as our planet undergoes a compression factor related to its mass, nearly twenty times higher. This implies that the proportion of heavy elements on Mercury is significantly higher than in other terrestrial planets. What happened to the lighter solid elements, which should logically be found closer to the surface? Maybe they were ejected following a major collision; the huge Caloris basin, dating back 3.8 billion years, demonstrates the possibility of such a hypothesis.

Another discovery is to the credit of the Mariner 10 spacecraft: the existence of a magnetic field detected by the presence of a magnetosphere. The magnetic field is weak compared with the Earth's magnetic field, but

The Mariner 10 spacecraft (© NASA).

its presence on Mercury is very unexpected. According to its high density, Mercury must have an iron-rich core, but given the low mass of the planet, we would expect a solid core; in addition, the low rotation of the planet is not expected to induce a magnetic field. Whatever the reason, the study and understanding of Mercury's magnetic field is among the major challenges in the exploration of Mercury.

Launched in 2007 by NASA, the Messenger mission made three flybys of Mercury between January 2008 and September 2009. The probe has been in orbit around the planet since March 2011 for an in-depth exploration of Mercury and its environment. The three flybys helped complete the longitude mapping of the surface; it covers now all latitudes except the poles. The images showed, in some areas, signs of increased tectonic activity (contraction or expansion of the crust). Much more surprising, crater counting highlighted regions that date to less than a billion years ago, which could be a sign of relatively recent volcanic activity. The morphology of some craters and the presence of some very flat plains,

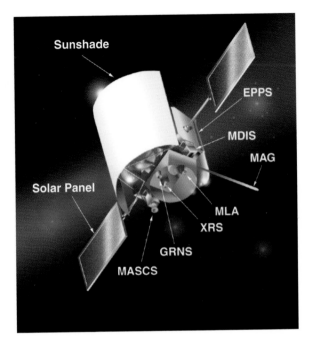

Figure 4.3. The Messenger probe (© NASA).

most likely filled with lava, support this hypothesis. This discovery has boosted the interest of planetologists who had long thought that planet Mercury had been inactive for billions of years. New results are expected from orbital observations by Messenger, and later from the European mission BepiColombo, currently under development at ESA, in partnership with the Japanese space agency JAXA; the launch of the mission is scheduled for 2015. Two orbiters will explore the planet simultaneously; the European orbiter will be dedicated to the study of its surface and its interior, while the Japanese orbiter will study the magnetosphere of Mercury and the interaction of the solar wind with the planet.

4.2 A Moon that looks like Mercury

If one considers the distance to the Sun, the Moon should not appear in this section, but after that of Venus. Moreover, it is not a planet, but a

Figure 4.4. The surface of Mercury, observed by the camera of the Messenger spacecraft (© NASA).

satellite. However, its physical nature justifies our ranking this satellite among the terrestrial planets, just after Mercury.

The Moon is indeed, like Mercury, a rocky body devoid of atmosphere: its gravitational field is insufficient to retain a stable atmosphere. In addition, its cratered surface bears a striking resemblance to that of Mercury. However, unlike Mercury, the Moon cannot be seen as a typical telluric object, which could have been discovered in a planetary system like ours, at the same distance from its star. As we have seen above (Section 3.8), the surprisingly low density of the Moon is the result of its formation process. Following the collision of Earth with a young proto-planet whose mass was comparable to that of Mars, the Moon has accreted, in Earth orbit, the lighter elements coming from the outer layers of the two bodies. We do not expect to find, at the orbital distance of the Earth, a rocky planet with such a low density.

We are also interested by the Moon for another reason. Whatever the reasons (far from being dominated by science), the Moon has been, with the American Apollo missions, the target of an unprecedented space

Figure 4.5. The surface of the Moon photographed by the Japanese probe Kaguya on October 31, 2007 from an altitude of 100 km (© JAXA).

exploration program. The result is a major scientific discovery: the absolute dating of lunar samples. Space exploration by the Apollo missions has allowed us to characterize the surface. After several decades of interruption, the Moon has become again a major target of interest for new space agencies: after Clementine and the technology mission SMART-1, launched respectively by NASA and ESA in 1994 and 2003, robotic missions orbiting the Moon have been sent or are under study by Chinese, Indian, Japanese and American space agencies.

The Moon's surface is divided into two major categories: the highlands, with a high albedo, and the plains, which are darker. The highlands, very cratered, are older and have a relatively low density (2.75 g/cm^3). Plains, or "seas" are of more recent formation. They are the result of large impacts and are filled with grey basaltic lava.There are some regions with strong vertical drops; near the south pole, the Leibnitz Mountains have an altitude of nearly 5 km. It is believed that the lunar mountains are the result of meteoritic bombardment and not, as on Earth, plate tectonics.

Lunar geology is well known from the Apollo and Clementine spacecraft. The materials that form the surface are essentially silicates, with the same composition as on Earth. These materials are covered with a layer of regolith, debris resulting from meteorite impacts, whose thickness can reach several meters. Basaltic lavas are enriched in iron as compared with the highlands and have a low viscosity, hence their fluidity.

In recent years, we are witnessing a new start to lunar robotic space exploration. Launched in October 2008 and operated until August 2009, the Chandrayaan 1 probe, launched by the Indian space agency ISRO, has caused a big surprise by announcing the detection of trace volatiles (water vapor and ice water), adsorbed on the lunar surface. These results were confirmed by other measurements, in particular those of the Cassini spacecraft, made during its flyby of Earth in 1999. Traces of moisture may be the result of the impact of the lunar surface by the solar wind, but the adsorption mechanism is still not well understood.

What will be the future of the lunar space exploration? Issues are primarily political but science could benefit from it. Projects are under study at ESA, in Japan and in the U.S.; so far France, more involved in the space exploration of the terrestrial and giant planets, has no strong involvement in them.

4.3   Venus, the furnace

Venus is the brightest planet in the sky. Visible, like Mercury, at sunrise and sunset, the "evening star" has been used as a celestial landmark since antiquity. At about 0.7 AU from the Sun, Venus is by its size and density very similar to the Earth. There is major difference, however: its rotation axis is close to perpendicular to the ecliptic, but it is facing south, opposite to all the other planets of the solar system. Its rotation is retrograde, with a very long rotation period of 243 days. What is the origin of this anomaly? It might be a collision early in its history, or chaotic dynamic evolution; the answer is not completely settled yet.

4.3.1 Traces of a past volcanism... and perhaps present?

The images of Venus in visible light immediately show that the planet is very different from Earth. An opaque and uniform cloud of greenish white color, located at pressure levels between 1 and 0.1 bar, hides the surface.

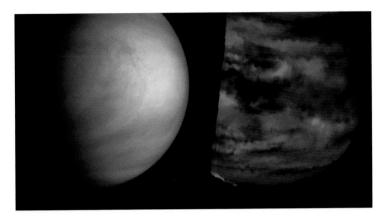

Figure 4.6. Planet Venus observed in ultraviolet (left) and infrared (right) by the Galileo space-craft during its flyby of the planet in December 1989 (© NASA).

With measurements from the ground made in the 1980s in the radio range, and thanks to the multiple spacecraft (Venera probes launched by the USSR, and Pioneer Venus and Magellan, launched by NASA), we now know the conditions of the hot lower atmosphere: the surface temperature is 90 bars and the temperature is 730 K. In 1982, the Soviet Venera probes sent us the first images of the ground of Venus. In 1992, the radar of the U.S. probe Magellan gave us a complete reconstruction of the surface of Venus. We know that Venus today has no plate tectonics, which could be due to the absence of water. The crust, of substantially uniform thickness, is made of basalt. In addition, the surface of Venus is young (less than a billion years), as evidenced by the small number of impact craters present on its surface; this young age is a sign of recent volcanism. No magnetic field was detected, and we have no information about the interior of Venus. The absence of magnetic field may be due to the fact that the core is solid, or to the absence of convection within a partially liquid nucleus.

The atmosphere of Venus is largely dominated by carbon dioxide (96.5%), with a small contribution of nitrogen (3.5%). Other minor constituents (CO, H_2O, SO_2, H_2S) are present in trace amounts, with abundances varying with altitude. Between 40 and 70 km altitude, the planet is covered with several cloud layers, all mainly made of sulfuric acid H_2SO_4. In the visible range, the clouds mask completely the lower atmosphere of

Figure 4.7. The surface of Venus photographed by Venera 13 (top) and Venera 14 (bottom) in March 1982. (© Academy of Sciences of the Soviet Union).

Venus. Fortunately, in the near infrared, there are partial spectral windows between the absorption bands of the carbon dioxide, wherein the radiation is emitted outwardly from the lowest layers of the atmosphere, and even, in some cases, from the surface itself. Spectroscopy in the near infrared thus provides a unique way to probe the lower atmosphere of Venus. Measurements were obtained in the early 1990s, from the ground and with the Galileo spacecraft as it flew over the planet before going to the Jupiter system, its final destination. All these measurements helped to reconstruct what is probably the sulfur cycle on Venus: at the clouds, SO_2, present in the lower atmosphere, is photolysed into SO_3, then reacts with H_2O to

Chapter 4. The Neighbors of the Earth

form sulfuric acid, H_2SO_4, which condenses and falls to the surface where it may form calcium sulfate $CaSO_4$. The permanent presence of sulfuric acid clouds suggests the existence of a source of SO_2, probably by volcanism. The images of the surface reconstructed from the Magellan radar data showed that the planet was covered with relatively young volcanoes, dating from the last billion years. However, no evidence of active volcanism has so far been detected on the surface of Venus.

Since 2006, the European spacecraft Venus Express has been operating in orbit around the planet, studying its atmosphere and its interaction

Figure 4.8. The Venus Express spacecraft. Launched in 2005 by ESA, the satellite has been operating in Venus orbit since 2006 (© ESA).

with the interplanetary medium. Venus has, like the Earth, a meridional atmospheric circulation made of so-called "Hadley" cells. At the equator, the gaseous masses heated by the Sun rise and move to higher latitudes where they come down. The system differs from the terrestrial circulation for two reasons: the very low rotation period and the lack of obliquity. As a result, on Venus, the cells rising from the equator reach high latitudes while, on Earth, they descend in the tropics to form two other systems at higher latitude. There is another unexpected phenomenon: observations of Venus have shown that the atmosphere is in super-rotation relative to its surface. At the equator, the zonal rotation period (i.e. along parallels) at the cloud top is only four days, while it is 245 days on the surface! The origin of this phenomenon could be the existence of thermal waves generated locally at the subsolar point (i.e. the point of the surface where the Sun is seen at zenith).

4.3.2 The past history of Venus as revealed by heavy water

Today, the water content in the Venus atmosphere is very small: less than one ten-thousandth of the total pressure in the lower atmosphere, even less above the clouds where it is trapped with SO_3 form H_2SO_4. However, we have evidence that water was present in large amounts in the early atmosphere of Venus. Where does this result come from? The answer is given by the HDO molecule, also called heavy water. In the heavy water molecule, one of the hydrogen atoms is replaced by its deuterium isotope, consisting of a proton and a neutron, and therefore twice as heavy as hydrogen. HDO presents spectroscopic signatures different from those of H_2O, and its abundance in a celestial body can be measured independently of that of normal water H_2O. Both species can be measured in the thermal infrared spectrum from the lower atmosphere of Venus. The striking result is that all measurements, including those made *in situ*, show a considerable enrichment of deuterium: the D/H ratio on Venus is about 120 times higher than on Earth! An even larger enrichment (up to 200) has been measured above the clouds by Venus Express.

To explain such a discrepancy, an explanation, now widely accepted by the scientific community, has been proposed. Water was probably once very abundant on Venus, perhaps as abundant as on Earth today. But it has

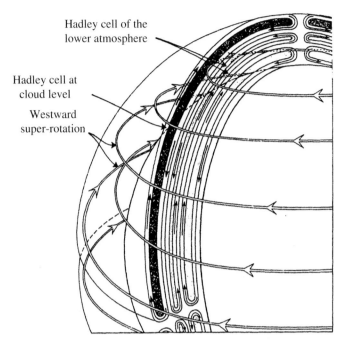

Hadley cell of the
lower atmosphere

Hadley cell at
cloud level

Westward
super-rotation

Figure 4.9.   The atmospheric circulation of Venus. As with the Earth, it is a Hadley-type convective circulation. Due to the very slow rotation of the planet, the cells starting from the equator expand up to the poles (after G. Schubert, in "Venus", D. M. Hunten et al., University of Arizona Press, 1983)

gradually dropped, probably after being destroyed by solar UV radiation, following a mechanism which is still poorly understood. The H_2O molecule, being lighter than HDO, escaped preferentially, resulting in a progressive enrichment of the atmospheric HDO residual.

Water escape may have been favored by the absence of a magnetosphere around Venus. It is generally believed that the magnetosphere, when it exists (this is the case of the Earth) acts as a protective cavity facing the flow of solar wind particles (see Section 3.6). But Venus, as we have seen, has no magnetic field, which is surprising given the conditions of mass and density, very similar to those of the Earth. Even taking into account the slow rotation of the planet, we would expect a field larger than the upper limits which were measured. Why is Venus devoid of a magnetic field? Probably because there is no motion in the convective core.

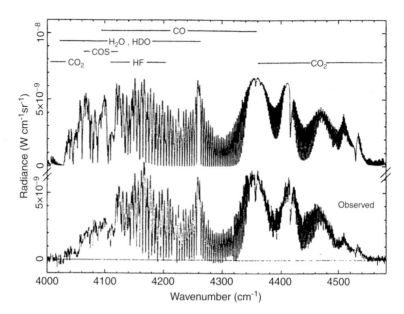

Figure 4.10. The thermal spectrum of Venus in the near infrared range. Top: synthetic spectrum; bottom: observed spectrum. Recorded on the night side of the planet, between strong absorption bands of CO_2, the thermal spectrum of Venus probes the deep troposphere below the clouds, and allows us, in particular, to measure the D/H ratio, from HDO/H_2O. After B. Bézard et al., *Nature* **345**, 508, 1990.

The cause is probably an insufficient cooling rate of the core, due itself to the absence of tectonic plates. Again, the lack of water has probably played a role in this situation. On Earth, it is believed that water promotes plate tectonics by its recycling in the mantle. The disappearance of water on Venus had the effect of inhibiting the processes that would have led to the initiation of this mechanism.

Let us return to the early history of Venus. Carbon dioxide and water vapor are powerful greenhouse gases. At a distance of 0.7 AU from the Sun, the water on Venus is found in the gas phase as soon as the Sun reached its present level of radiation. Note that this was not the case at the beginning of its history: the models of stellar evolution tells us that the young Sun emitted around 70% of its current energy. At this time, the water of Venus may well have been in the form of an ocean. If so, it must have gradually evaporated as the solar flux increased. Nothing could stop the greenhouse effect on Venus, which is responsible for the current furnace.

Venus Express continues its exploration of the atmosphere, its dynamics, its interaction with the surface and with the outside world. Other projects are under consideration in Europe, the United States and Russia. The main objectives of this research are, in particular, to determine whether Venus now has an active volcanism, and to understand the reason for the absence of magnetic field and plate tectonics. It is unfortunately very difficult to obtain information about the distant past of the planet, since the relatively recent remodeling of the surface by volcanic activity has erased any indication prior to the last billion years.

4.4 Mars, a desert word

With Mars, the red planet, we move beyond the Earth, into the world of the outer planets. Compared with Venus and the Earth, Mars is both smaller and less dense; its mass is only one tenth of that of the Earth. Its atmosphere, as we have seen, is extremely tenuous, so that its surface is directly observable with a ground-based telescope. While Mercury and Venus are without satellites, Mars has two very small ones (less than 30 km in size), Phobos and Deimos (see Section 3.8). They are probably asteroids captured by the planet.

From the beginning of the space adventure, Mars was a prime target: the first attempts, American and Soviet, date from the 1960s. The history of Mars' exploration is a journey marked with failures: the loss rate was very high initially and remains, after forty years, around 50%. The first success came from American missions Mariner and Viking. The first main step was Mariner 9 in 1972, which provided us with the initial mapping of the surface, revealing volcanoes and canyons. This major discovery put an end to the controversy which arose from observations of Schiaparelli a century ago, and lasted for decades, supported in particular by the American astronomer Percival Lowell, concerning the possible existence of "channels" and traces of life on Mars. The exploration of Mars, led by NASA, was made with orbiters and descent modules, and then, since 1997, with mobile devices. Since 2003, Europe has in the adventure with

Figure 4.11. Planet Mars as observed by the camera of the Viking probe (© NASA).

the Mars Express orbiter. The future of participated space exploration of Mars will most likely be based on an international collaboration between space agencies, with the prospect of returning Martian samples to Earth.

Despite its low mass, the planet Mars shows two remarkable similarities with the Earth: its rotation period and obliquity, both very close to terrestrial values. It follows a seasonal pattern that is reminiscent of ours, with the presence of polar caps, both seasonal and perennial. Another striking similarity is the relief of Mars, characterized by volcanoes and canyons much larger than Earth, and deserts similar to ours.

The similarity stops here. There is a major difference in the seasonal cycles of Mars and Earth, linked to both the atmospheric composition of Mars, dominated by carbon dioxide, and its low temperature. A periodic condensation of carbon dioxide takes place, with a deposition of CO_2 ice on the north and south poles alternately, depending on the seasonal cycle: one third of the atmosphere is trapped at the poles in this process, which leads to very strong winds and dust storms that can cover the entire planet. In addition to the carbon dioxide cycle, there are two other cycles associated with water and dust. Although water vapor is a very minor constituent of the Martian atmosphere with a volume ratio of less than 1%, water, like carbon dioxide, alternately condenses at the poles (directly from the gas phase to the solid phase) and contributes to the seasonal caps. The dust cycle is associated with the strong temperature contrast which is observed during the southern summer as a function of the large eccentricity of Mars.

Let us go back to the surface of Mars. It carries the largest known volcanoes in the solar system: the highest, Olympus Mons, 25 km high, is at the limit allowed by isostasy (see Section 3.3). The huge canyon Valles Marineris, 4500 km long, 150 km wide and 8 km deep in places (the analogue, albeit much bigger, of the great African Rift on Earth), also marks a very active tectonic activity in the past. Unlike Venus, Mars retains traces of its ancient past, until about 3.8 billion years ago in the heavily cratered, southern highlands. In the southern hemisphere is found the huge Hellas impact basin, whose depth reaches more than 5 km below the average level. In the northern part, the low plains have been covered with lava.

4.4.1 A thin and dry atmosphere

The atmospheric composition of Mars, with 95% CO_2, 3% N_2 and 1% Argon, is strikingly similar to that of Venus. Other minor components have mixing ratios below the percent: O_2 (0.13%), CO and H_2O (about 10^{-4}), then O_3, H_2O_2,... (about 10^{-8}). Unlike Venus, we do not find any sulfur products.

With an average surface pressure of 6 mbar, the Martian atmosphere is very responsive to changes in solar flux, which leads to strong temperature differences between summer at the equator (about 0°C or more) and winter at the poles (−100°C). Periodic condensation of CO_2 and H_2O at the poles of Mars, added to the low atmospheric mass, leads to very complex atmospheric dynamics, increased by the eccentricity of the Martian orbit, higher than that of the Earth, which enhances seasonal contrasts.

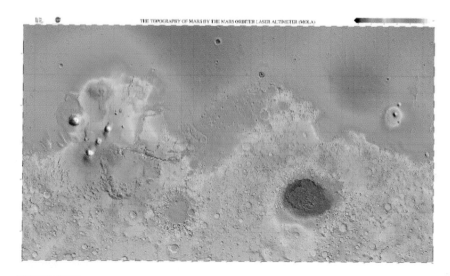

Figure 4.12. The altimetry of Mars, determined by the laser altimeter of the Mars Global Surveyor (© NASA).

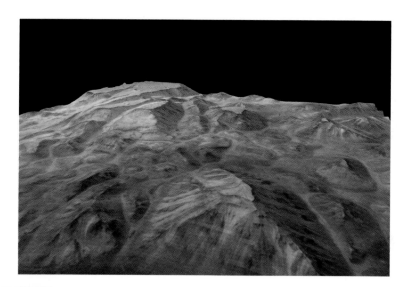

Figure 4.13. The canyon Valles Marineris observed by the HRSC camera of the Mars Express orbiter. With a length of 4000 km, a width of 200 km and a depth of 7 km, Valles Marineris is a huge canyon at the surface of Mars, over ten times larger than the Grand Canyon of Colorado (© ESA).

Chapter 4. The Neighbors of the Earth

As on the Earth and Venus, the Martian atmospheric circulation is convective and characterized by Hadley cells. At the equinox, the heated gas rises at the equator and descends at higher altitudes; at solstice, the cell rises from the subsolar point (i.e. the point at which the sun is zenith) to cross the planet and falls down at opposite latitudes. The perihelion occurs during the summer in the southern hemisphere, where the temperature is maximum, which gives rise to particularly violent dust storms. Strong winds of up to 100 km/h on the ground and 400 km/h at 50 km altitude generate these storms which can cover the entire planet and last for several months.

4.4.2 Water on Mars: Past and present

Pressure conditions now are such that water cannot be in liquid form at the surface. However, many indications seem to suggest the presence of liquid water in the past history of Mars. But when was the water liquid, and how

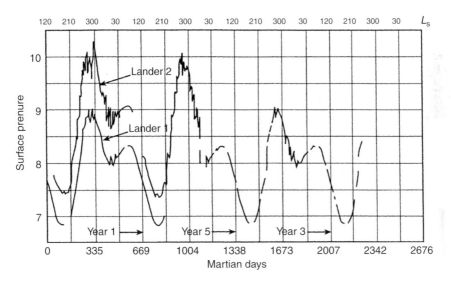

Figure 4.14. Pressure variations at the surface of Mars recorded as a function of the season, measured *in situ* by the Viking landers over three Martian years, i.e. about six Earth years. The seasonal modulation of pressure is the result of the condensation of carbon dioxide at the poles in winter. A Martian day corresponds to about 2 terrestrial days (© NASA).

long did it stay? This will be the main challenge of the exploration of Mars in the coming decades.

With a mixing ratio on the order of 10^{-4}, the water vapor on Mars today is a very minor component of the Martian atmosphere. Its abundance, integrated over the atmosphere, varies from 0 (at the pole in winter) to 100 precipitable-microns (around the north pole in summer), depending on the condensation of H_2O into ice. It is currently difficult to make an overall assessment of the total amount of water on Mars. It seems clear that water ice is present below the Martian polar caps. Measurements made by the Mars Odyssey in 2000 detected the presence of hydrogen atoms below the poles; radar and infrared measurements made by Mars Express and Mars Reconnaissance Orbiter have confirmed this result. However, the abundance of water remains to be determined. Water can also be trapped below the surface at other latitudes, in the form of permafrost. In fact, lobate ejecta craters, present mostly at high latitudes, seem to be indicative of the past presence of water in the subsurface.

What are the signs for the past presence of liquid water on Mars? They go back to the first discoveries of Mariner 9 and the Viking missions, whose images revealed surprising structures: on the one hand, valley networks, now drained, in old terrain and, on the other hand, "outflow valleys" corresponding to more recent catastrophic collapses. More recently, infrared measurements of Mars Express and *in situ* observations of the Opportunity rover showed the presence of sulfates, which, on Earth, can only be formed in the presence of liquid water. Sulfates on Mars date back to relatively recent times (–2 to –3 billion years). Liquid water on Mars would have existed in that period, but for a limited time: indeed the Spirit robot did not find any sedimentary deposits that would have been expected if the water had stayed on the surface for a long time. On the other hand, the presence of hydrated silicates or clays (phyllosilicates) in some ancient rocks of the southern hemisphere suggests that liquid water existed in an earlier period, at the time of the formation of these terrain (over 3.5 billion years). Carbonates have been searched for, as they could have been trapped in the Martian soil in the presence of liquid water (as was the case on Earth), but only traces have been found. The formation of carbonates may have been inhibited in the presence of sulfur dioxide

Figure 4.15. The surface of Mars as observed by the Viking lander (© NASA).

emitted by volcanoes. But SO_2 has not been detected either in the Martian atmosphere.

Our understanding of the history of water on Mars is still very incomplete. We can try to summarize it as follows: liquid water flowed at the beginning of the planet's history (over 3.5 billion years ago), as evidenced by the valley networks and the presence of hydrated silicates in some of these sites. Water then disappeared but continued to flow sporadically during local phenomena (perhaps associated with volcanic or tectonic activity) that was responsible for the outflow channels and the formation of sulfates. But this is only a hypothesis that remains to be validated. A few years ago, some researchers raised the possibility of an ancient water ocean that would have originally covered the northern plains, and recent radar measurements support this hypothesis. The debate is ongoing.

4.4.3 A hot and humid climate in the past?

It seems certain that liquid water flowed on Mars in the past. The implications are considerable: for the liquid water to stay on the surface, it is necessary that the pressure and temperature be significantly higher than today's. We have two clues that reinforce this hypothesis.

The first clue comes from measurements of isotope ratios in the atmosphere of Mars. The first of them is the D/H ratio. We have seen that in the case of Venus, the high value of the HDO/H_2O ratio (120 times the terrestrial value) has been interpreted as evidence for an escape of a large amount of water vapor in its history. In the case of Mars, a deuterium-enrichment was also measured by infrared observations from Earth.

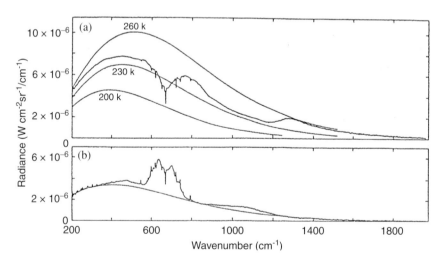

Figure 4.16. The thermal spectrum of Mars recorded at the equator (top) and at the pole (bottom). At the equator, the surface temperature is greater than that of the atmosphere and the band of carbon dioxide, at 15 μm, appears in absorption. At the pole, the surface temperature is lower than the atmospheric temperature and the CO_2 band appears in emission. After R. Hanel *et al.* Infrared Remote Sensing of the Solar System, CUP, 1992.

Chapter 4. The Neighbors of the Earth

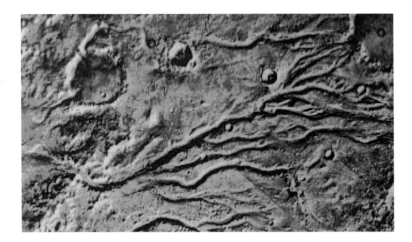

Figure 4.17. Example of valley network on Mars (Valles Vedra and Maumee, near Chryse Planitia, (© NASA).

The D/H ratio, derived from HDO/H_2O, is enriched by a factor of 5 with respect to the terrestrial value. It implies that on Mars too part of the atmosphere escaped. The same conclusion is reached with the $^{15}N/^{14}N$ isotopic ratio measurements obtained with the Viking probes. The enrichment by a factor of 1.6, as measured with respect to the terrestrial value, implies also a significant outgassing of the atmosphere. According to some theoretical models, the primordial atmosphere of Mars could have been ten times denser than today's.

How could the atmosphere escape? This is where the second diagnostic is found. In 1999, the Mars Global Surveyor spacecraft revealed the existence of a fossil magnetic field, trapped in ancient rocks in the southern hemisphere. The remnent magnetization of these lands is interpreted as the existence of a dynamo at the time of their formation during the first billion years of the history of the planet. This dynamo eventually stopped due to the lack of sufficient internal energy (let us remember that Mars is ten times less massive than the Earth). But during the phase of internal activity, Mars possessed a magnetosphere that could prevent the air from escaping. After the generator turned off and the magnetosphere disappeared, Mars may have gradually lost its atmosphere; in parallel, the greenhouse effect was reduced and the temperature decreased until it reached the values that we know today. In another scenario, the planet could have experienced a giant meteorite impact (or more), which would have resulted in a sudden and massive outgassing of most of its atmosphere. It is not currently possible to decide between these two hypotheses.

Beyond this debate emerges a question that opens another dimension. We now know that liquid water existed on Mars in the past. Could it have stayed long enough for life to appear there? If yes, can we hope to discover traces of fossil or even present life? This question is at the heart of the future exploration of Mars. A recent discovery came to stimulate the discussion even further. The detection of methane was reported on Mars in 2004, in very small amounts (with a mixing ratio of the order of 10^{-8}), on the basis of different space and ground-based observations. Methane emissions seemed to be localized and variable. If real, they would imply the existence of sources and sinks not yet identified. On Earth, methane is almost entirely of biogenic origin, with a few abiotic sources related to geothermal activity. On Mars, such a source could also be considered. It would mean that, under the surface, reactions between silicate and liquid water are forming hydrogen, which then reacts with carbon dioxide to form methane. This mechanism would therefore require the presence of an aquifer (a mixture of rocks and liquid water at high pressure) beneath the surface of Mars. Its presence was discussed on the basis of geophysical models but has not been demonstrated. However, at the end of 2012, the Curiosity rover searched for methane and could not find any trace of it. Was the 2003 detection of CH_4 real? If so, was it a single event? Again, the problem remains.

4.4.4 From the distant past to the recent past...

A curious discovery has recently demonstrated the power of numerical simulation methods for reconstructing the orbital characteristics of the planets, and their possible implications on the planets' past climates. Such work has highlighted the periodic evolution of the inclination and the eccentricity of Mars over the last tens of millions of years. In particular, the obliquity of the planet is now very close to that of the Earth (24°), but

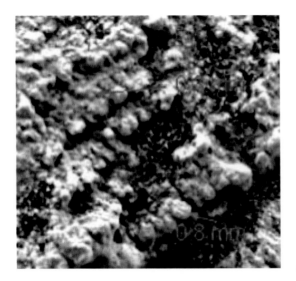

Figure 4.18. These grains observed by the Opportunity rover are probably spherules sulfates. On Earth, these materials are formed in the presence of liquid water salt, which could imply its presence on the surface of Mars in the past history of the planet (© NASA/JPL/Cornell).

could have reached 40° twenty million years ago. Climate evolution models show that water ice, now trapped at the poles, was then in the form of equatorial glaciers, especially on the flanks of reliefs. Images by Mars Express showing morphological residues of these glaciers, at the foot of volcanoes and on the side of the Hellas Basin, brought a dramatic confirmation of this theory. Incidentally, the discovery of climate changes over the long term associated with orbital variations reminds us of the role that the Moon probably played in the evolution of Earth's climate by stabilizing the obliquity of the planet.

4.5 Comparative evolution of terrestrial planets: the role of water

In light of the previous discussions, we can better understand how and why the terrestrial planets, which started from relatively close initial conditions, have evolved to their present state.

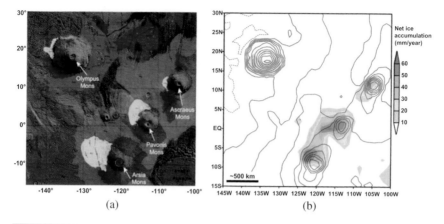

Figure 4.19. Glaciers on the flanks of Martian volcanoes: left, observations with the HRMC camera of the Mars Express orbiter (© ESA); right, predicted locations from numerical simulations based on the obliquity variations and Global Climate Model (after F. Forget *et al.*, 2006)..

Let us first compare Venus and Earth. Their sizes and densities are very similar: they really seem like twin planets. The main difference is their heliocentric distance: the solar flux received by Venus is twice that received by Earth, and the equilibrium temperature at the orbit of Venus, assuming an equal albedo for both planets, is higher by about 10%. It is sufficient for the water to be, similar to carbon dioxide, in the form of gas. Even though the water on Venus could have been in liquid form at the beginning of its history when the Sun was less bright than today, this phase could not keep up with the increasing solar flux. The presence of two very effective and abundant greenhouse gases, H_2O and CO_2, was the cause of the runaway greenhouse effect, responsible for the high temperatures observed today. The water gradually disappeared by photodissociation and escaped, and the atmosphere of Venus became mainly composed of CO_2, with a minor contribution from N_2.

Earth, however, is found at a heliocentric distance such that the surface temperature is between 0°C and 100°C, so the water can remain in liquid form. Carbon dioxide, very abundant in the early atmosphere, was trapped on the ocean floor by reaction with rocks to form carbonates, mainly calcium carbonate ($CaCO_3$). Nitrogen is spectroscopically inactive and therefore has no effect on the greenhouse effect; the greenhouse effect

Chapter 4. The Neighbors of the Earth

was thus very limited, hence the temperature remained approximately constant over time, with the exception of certain phases of glaciation, probably interrupted by volcanic episodes. The emergence of life, nearly four billion years ago, led to the formation of oxygen and the ozone layer. This made possible the development of life on the continent about 500 million years ago.

As for Mars, it differs from Venus and the Earth in two ways: firstly, it is about ten times less massive, and secondly it is farther from the Sun, so the solar energy received at the surface is about two times less than the Earth and four times less than that of Venus. The low mass of Mars has the effect of a lower internal energy, as internal energy comes from the radiation of long-lived radioactive elements contained in the core. If the amount of these elements is low, the radiation decreases and the planet becomes inactive. The extinction of the dynamo of Mars before the end of the first billion years seems to confirm this scenario. Another consequence of the low mass of Mars is that its gravitational field is less able to capture a massive atmosphere. In the beginning, the Martian atmosphere was close in composition to that of its neighbors, but it must have been significantly less dense.

Under current conditions, water is mostly in the gaseous phase on Venus, and liquid and solid on the Earth and on Mars. But this has not always been the case since, as we have seen, liquid water flowed on the surface of Mars in the past, and it may also have been present at the beginning of the history of Venus. In the case of Mars, the insufficient amount of internal energy probably resulted in the extinction of the dynamo, the disappearance of the magnetic field and perhaps also the escape of the atmosphere.

What can we learn from the comparative evolution of the terrestrial planets? Even in the presence of comparable initial conditions, planets can evolve in a radically different way; water, with its phase changes, plays a major role in these developments; the masses of the planets are drives of their internal activity; and the presence of a magnetosphere can probably slow down or even prevent the atmospheric escape. When we look at the diversity of terrestrial planets, we can guess that we will find an even greater diversity among the extrasolar "exo-Earths" that the instruments of the future will discover around other stars in the decades to come.

5

A Little Further, the Giant Planets

Beyond the snow line, more than 5 AU from the Sun, lies the realm of the giant planets. We have seen (Chapter 2) how the mode of formation of planets from solid particles in a rotating disk had favored the emergence of two distinct classes of planets, the terrestrial and the giant ones. Giant planets are formed in two stages: first from the accumulation of an ice core of about ten Earth masses, followed by the collapse of the surrounding gas (mainly consisting of hydrogen and helium), captured by the gravitational field of the core. The main characteristics of the giant planets result from this formation process: they are very large (the radius of Jupiter is eleven times that of the Earth, those of Uranus and Neptune four times), but they also have a very low density (0.7–1.6 g/cm^3). Giant planets share another characteristic: they are all surrounded by a ring system and a system of satellites. Many of them are regular, i.e. they are located, like the rings, in the equatorial plane of the planet around which they revolve. Their presence is a consequence of the collapse of the gas surrounding the nucleus of ice around it; we can consider the giant planets and their systems as small miniature solar systems. There is still a limit to this analogy: compared to the giant planets they surround, the outer satellites are much closer and larger than the planets are from the Sun (Io, the closest Galilean

Figure 5.1. Planet Jupiter, observed by the camera of the Cassini spacecraft during its flyby in December 2000 (© NASA).

satellite of Jupiter, is located at 6 Jovian radii from the planet, while Mercury, the closest planet to the Sun is at more than 80 solar radii from the Sun). Note also that there is no ring system in the immediate vicinity of the Sun.

5.1 Two classes of giant planets

5.1.1 Gas giants and icy giants

Let us look closer at the four giant planets. There is Jupiter first, closest to the Sun, and also the largest and most massive: its mass is equal to 318 Earth masses. Now located at 5.2 AU from the Sun, it was formed just after the snow line, and thus has benefited from a large amount of solid material made available by the ice condensation. With its yellow–orange parallel band and zones structure and its Great Red Spot, identified by Galileo over three centuries ago, Jupiter has always been an object of attention for astronomers. About twice as far, at 9.2 AU from the Sun, we find Saturn and its impressive ring system, known for over three centuries. Huygens gave, in the seventeenth century, the explanation of the nature of

Figure 5.2. Planet Saturn observed by the camera of the Cassini spacecraft (© NASA).

the rings, whose appearance changes periodically, depending on the position of the Earth relative to its plane. With 95 Earth masses, Saturn, like Jupiter, is mainly made of gas from the initial nebula (let us remember that the mass of the icy core is only about ten Earth masses). This is why Jupiter and Saturn are called "gas giants". Located farther away, at 19 AU from the Sun, we find Uranus, then, at about 30 AU, Neptune. These two planets, very different from Jupiter and Saturn, are similar in mass, radius and density. With masses of 14 and 17 Earth masses, they are composed mainly of ice: they are called "icy giants".

What is the origin of these two categories in the giant planets? Here is a possible hypothesis: Jupiter and Saturn, being closer to the Sun, benefited from a larger mass of ice to form their initial core; this core could reach the critical mass of 10 Earth masses before the T-Tauri phase of the

Figure 5.3. Planet Uranus observed by the Voyager 2 spacecraft in 1986 (© NASA).

Chapter 5. A Little Further, the Giant Planets

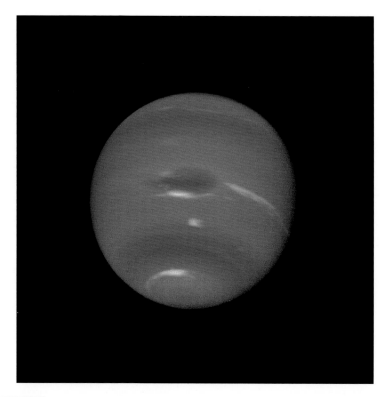

Figure 5.4. Planet Neptune photographed by the Voyager 2 spacecraft in 1989 (© NASA).

young Sun had the effect of dispersing the surrounding gas, about ten million years after its formation. In contrast, Uranus and Neptune, farther outside in the protoplanetary disk, with less solid material available, needed more time to build their initial nucleus. It is quite possible that the critical mass was reached after the dispersion of the gas. The planets could then accrete only a small contribution of gas.

5.1.2 The chemical composition of giant planets, a test of their formation scenario

The whole formation scenario of the giant planets is built, as we have seen, on the assumption of an initial ice core which accretes the gas of the protosolar environment. According to theoretical models and numerical

Figure 5.5. Diagram of the giant planets, illustrating the internal structures of the gas giants (Jupiter and Saturn), mostly made of gas, and the icy giants (Uranus and Neptune) which mainly consist of their initial core of ice (not to scale).

simulations, the collapse of the gas occurs when the core mass reaches ten or fifteen terrestrial masses (see Section 2.5). This model has been adopted above to define two types of giant planets, the gaseous and the icy ones. But do we have evidence for the existence of ice cores in giant planets? The answer is yes, and it is provided by measuring the abundance of their elements. This is what we will detail below.

Within the nebula, the abundance of different chemical elements is a direct result of how they were formed: by primordial nucleosynthesis in the case of hydrogen, helium and deuterium, and by stellar nucleosynthesis, in stars, for all the other elements (carbon, nitrogen, oxygen, etc.) that we call "heavy elements". The abundance of elements in the

Chapter 5. A Little Further, the Giant Planets

protosolar disk reflects the so-called cosmic abundances. Hydrogen, with 75% of the total mass, is predominant. The second one is helium with nearly 23%. All other "heavy" elements are contained in less than 2% of the total mass — more accurately, 1.6%. What are these heavy elements? The main elements C, N and O are found in the state of ice (H_2O, CH_4, NH_3, H_2S ...) in the initial core of the giant planets. Let us evaluate now the amount of heavy elements in each giant planet, assuming an initial core of 12 Earth masses (i.e. the limit beyond which the surrounding gas can be captured). We assume that the core consists entirely of heavy elements, whose mass fraction is 1.6% of the total. We add the contribution coming from the surrounding protosolar gas, with the cosmic abundances.

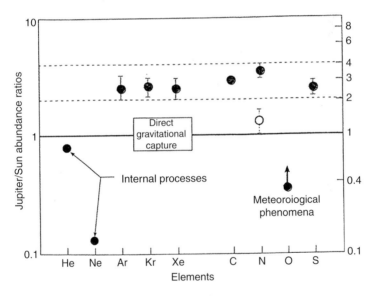

Figure 5.6. Enrichment of Jupiter in heavy elements as compared to the protosolar values. Abundances are compared with that of hydrogen. It can be seen that the six elements (C, N, S, Ar, Kr, Xe) show a enrichment between 2 and 4 times the protosolar value. The abundances of two elements, neon and oxygen, are significantly reduced. Neon is probably in condensed form inside the planet; regarding oxygen, the measurement is not representative of the global planetary value, due to the effects of atmospheric convection (the measurement was performed in a very dry subsidence region). Following a new determination of solar abundances, the average enrichment relative to hydrogen, equal to 3 +/− 1 in this figure, has been re-assessed to 4 +/− 2. The figure is taken from Owen *et al. Nature*, 1999.

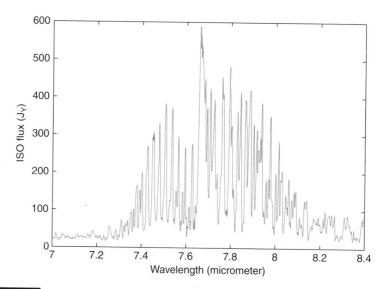

Figure 5.7. The spectrum of Saturn in the emission band of methane at 7.7 μm. The analysis of this spectrum led to a determination of C/H in Saturn's atmosphere, equal to 9 times the protosolar value. The spectrum was recorded by the Short Wavelength Spectrometer (SWS) instrument aboard the Infrared Space Observatory (ISO) satellite.

We then assume that, following the collapse of the protosolar gas and the initial core temperature rise that followed, the entire planet is re-homogenized; what is measured in the outer envelope is then representative of the entire planet. A simple calculation then shows that we should observe an enrichment in heavy elements with respect to hydrogen, as compared with cosmic abundances. This enrichment increases as the mass fraction of the initial nucleus is larger compared to the total mass of the planet. Thus we obtain, relative to cosmic values and with respect to hydrogen, an enrichment of 4 for Jupiter, 9 for Saturn and 30 to 50 for Uranus and Neptune.

Actually, these enhancements have been measured in the giant planets. In the case of Jupiter, the abundances of several elements were measured by the Galileo descent probe (we will come back to it). In the case of the three other planets, we only have measurements of methane which gives us access to C/H. In all three cases, the measurements are in good agreement with predictions. Further confirmation is provided by the D/H ratio measured from the ratio HD/H_2 and also CH_3D/CH_4. Again, all measurements are consistent: the D/H ratio is significantly higher on Uranus and Neptune (5×10^{-5}) than the protosolar value (about 2×10^{-5}). We know that D/H is enriched in ices (such enrichment is measured in the laboratory and in the interstellar medium) . Ices are especially abundant in Uranus and Neptune as their cores constitute the major fraction of their mass. In contrast, on Jupiter and Saturn, the D/H ratio is close to the protosolar value. So here is a simple diagnostic to support the nuclear formation model of the giant planets.

5.1.3 Initially, a moderate migration

Let us observe again Uranus and Neptune: there is still an intriguing fact. Why is Neptune, now located about 30 AU from the Sun, more massive than Uranus, which is only at about 20 AU? *A priori*, one would expect that the density of matter in the protosolar disk decreases with the heliocentric distance. Why do we see the opposite?

There is a possible explanation for this paradox, but it is only a hypothesis. For twenty years, numerical simulations on the dynamics of solar system objects, and in particular their formation scenario, has been made possible by the commissioning of supercomputers. It is now possible to simulate the evolution of terrestrial and giant planets, as well as asteroids, from a choice of initial conditions. The initial conditions have to be selected in order to produce results closest to the current observations. In the case of giant planets, there is a general agreement on the fact that they have experienced in their history some variation in their distance from the Sun; this is called migration. This is a relatively new idea: it developed over the past fifteen years, following the unexpected discovery of giant exoplanets close to their star. To explain this phenomenon, in total contradiction with what we see in the solar system, it was necessary to invoke the migration process, which brings a giant planet originally formed away from the central star closer to it. If it appears that this mechanism is common in exoplanetary systems, could it have been at work also in the solar system? According to the theory developed by dynamicists at the Observatoire de Nice — wisely called "the Nice Model" — Jupiter, initially located about 6 AU from the Sun, would have moved slightly toward the inner solar system to reach its current position of 5.2 AU, while the orbit of Saturn would moved from about 8 AU to its current position of 9.2 AU. Uranus and Neptune, both initially located at about 15 AU from the Sun, have also experienced significant outward migration and, in some models, Neptune was originally closer to the Sun than Uranus. This is not the end of the story, however; more recent simulations suggest that, in an earlier stage, Jupiter approached the inner solar system down to the orbit of Mars (thus preventing the planet from growing further) and then moved back toward the outer solar system under the effect of Saturn's attraction.

The scenario of migration of the giant planets, although not firmly proven, has several elements in its favor. In addition to the relative sizes of Uranus and Neptune, it provides a plausible explanation for the phenomenon known as "Late Heavy Bombardment" (LHB, see chapter 2.3.3). We know, from the analysis of impact craters on the surfaces of solar system objects devoid of atmosphere, that a maximum impact rate occurred about 800 million years after the formation of the Sun and planets. It is quite possible that this phenomenon was initiated by the crossover of the 2:1 resonance of the Jupiter–Saturn system (Jupiter completing two revolutions while Saturn completes exactly one). Numerical simulations show that this condition leads to a great disturbance in the orbits of small bodies, with very high values of the inclinations and eccentricities, resulting in a very strong increase of the collision rate. Finally, the early motion of Jupiter inside the inner solar system would explain why Mars is ten times less massive than the Earth.

5.1.4 A hydrogen-rich atmosphere

Following the collapse of the protoplanetary gas around the initial icy nucleus, it is expected that the atmosphere of the giant planets — which is accessible to observation down to pressures of about ten bars — is dominated by hydrogen. This is indeed the case: molecular hydrogen, detected by spectroscopy in the early sixties, is the main atmospheric constituent (between 75% and 85% by volume). The rest consists mainly of helium, the other gases being only trace species. The most abundant of these minor constituents is methane, which represents less than 1% of the atmospheres of Jupiter and Saturn, and about 2% of Uranus and Neptune.

Methane (CH_4) is nevertheless important for several reasons. Firstly, it has a very rich infrared spectrum: it is, with CO_2 and H_2O, the most active greenhouse gas. In addition, it is photodissociated by solar ultraviolet radiation, which causes the production of several hydrocarbons (C_2H_2, C_2H_6, C_2H_4, CH_3, C_6H_6 ...). These gases, in turn, absorb solar energy causing the warming of the upper atmosphere and the presence of a stratosphere. The phenomenon is analogous to the formation, on the Earth, of the ozone layer and the stratosphere.

Unlike the terrestrial planets, the atmospheres of the giant planets are reductive. Elements are naturally associated with the hydrogen to form CH_4, NH_3, H_2O, H_2S .... Some condense, giving rise to clouds or haze in the stratosphere. In Jupiter and Saturn, NH_3 combine with H_2S and H_2O to form ammonium thiosulphate NH_4SH and ammonium hydroxide NH_4OH, which both condense into clouds; in the case of Jupiter, the cloud appears at about 2 bars. At lower altitudes (and therefore higher pressures), the

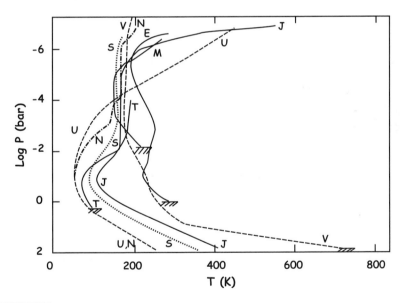

Figure 5.8. The thermal profiles of the planets and Titan (J: Jupiter S: Saturn U: Uranus, N: Neptune T: Titan, V: Venus E: Earth M: Mars). It can be seen that the four giant planets have close tropospheric gradients, and a tropopause located around 0.1 bar. The stratospheres, however, are very different, reflecting the different heating mechanisms.

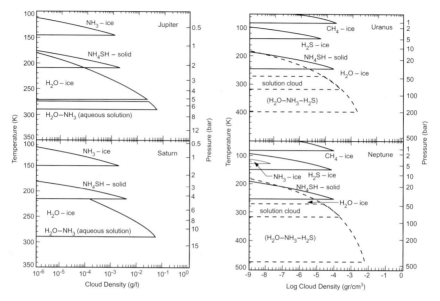

Figure 5.9. The cloud structure of the giant planets. On Jupiter and Saturn, ammonia NH_3 condenses around 0.5 bar, then NH_4SH around 2 bars and H_2O at a few bars. On Uranus and Neptune, which are colder, these species condense at greater depths, at levels inaccessible to observation. Methane condenses around 1 bar, and hydrocarbons (especially ethane C_2H_6), produced by the methane photodissociation, condense in the stratosphere. After S.K. Atreya and P. Romani, Planetary Meteorology (G. Hunt, edt. CUP) for Jupiter and Saturn, and after I. Pater *et al.*, *Icarus* **91**,220,1991 for Uranus and Neptune.

water condenses. For Jupiter and Saturn, the water cloud is located at a pressure of 5 to 10 bar. At higher altitude of the four giant planets, stratospheric haze may form, related to the condensation of hydrocarbons, particularly ethane C_2H_6.

The thermal structure is remarkably similar in the four giant planets, with a convective troposphere where, as on Earth, the thermal gradient is close to the adiabatic value. As the adiabatic gradient depends on the atmospheric composition (i.e. mostly hydrogen and helium), temperature profiles, depending on altitude, show comparable slopes on the four planets, with values decreasing as the heliocentric distance increases. A minimum occurs at the tropopause, at a pressure of about 100 millibars. Then the temperature profiles increase with altitude again, and the profiles are very

Chapter 5. A Little Further, the Giant Planets

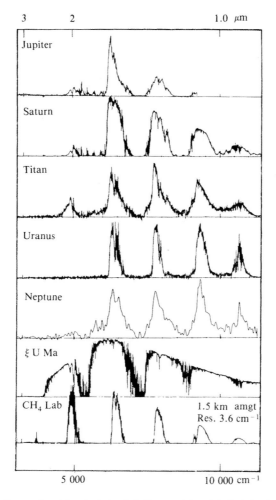

Figure 5.10. The spectra of the giant planets in the near infrared (reflected solar radiation) . The whole spectral range is dominated by methane absorption, as shown by its laboratory spectrum at the bottom. Although methane is a very minor atmospheric constituent, it also dominates the spectrum of the four giant planets and Titan. According to H. P. Larson, *Ann. Rev. Astron. Astrophys.*, 1980.

different from one planet to another. In all cases, the presence of hydrocarbons from the photodissociation of methane causes a rise in temperature and therefore a stratosphere. At higher altitudes, there are other factors: gravity waves, heating by high-energy particles brought by the magnetosphere, and more.

The spectrum of the giant planets is dominated by the spectral signatures of hydrogen and methane. The latter, although very small in the atmospheric composition, is present everywhere in the infrared spectrum, from the near infrared to the thermal range. Ammonia (NH_3) is clearly visible in the spectrum of Jupiter; because it condenses at lower temperatures, it is only weakly detectable on Saturn, and completely absent from the spectra of Uranus and Neptune. On Saturn, in contrast, phosphine (PH_3), which is expected to be less abundant in terms of cosmic abundances, is detected in abundance. This, anomaly has to be related to a more active vertical circulation in the case of Saturn. On Uranus and Neptune, most of the minor constituents are in the form of ice at pressure levels accessible to observation. It was therefore expected that the spectrum of the two icy giants would be limited to signatures of methane and hydrocarbons. In 1992, the unexpected

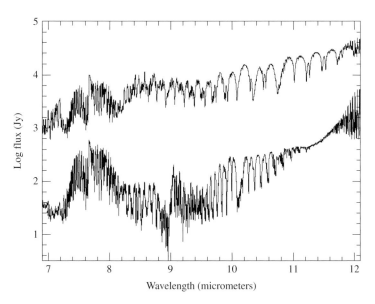

Figure 5.11. The thermal spectrum of Jupiter and Saturn between 6 and 12 μm, recorded by the spectrometer SWS ISO satellite. The spectra show a mixture of emission lines (CH_4 and C_2H_6 formed in the stratosphere of the planets at 7.7 and 12 μm respectively), and absorption lines (NH_3 and PH_3 around 10 μm; formed in the troposphere). The contrast between the spectra of these two planets, *a priori* similar, is striking. It can be explained by the lower temperature of Saturn (which causes condensation of NH_3), and by the active vertical circulation on Saturn (which leads to the presence of PH_3 and more intense stratospheric emissions). After T. Encrenaz *et al.*, Plan. and Space Science, 2003.

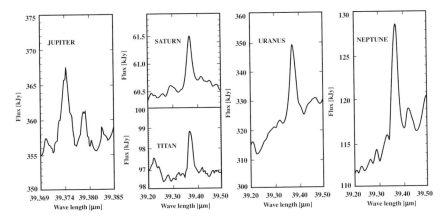

Figure 5.12. The emission spectrum of water vapor in the giant planets. The presence of water in the stratospheres of the giant planets and Titan is the sign of the presence of a flow of oxygenized material, of interplanetary origin (comets) or local origin (rings and satellites). After Lellouch *et al.*, Astron. Astrophys. 2002. (© T. Encrenaz, Searching for water in the Universe, Belin 2004).

detection CO and HCN in the stratosphere of Neptune was a surprise. The mixing ratio of CO was found to be a thousand times greater than its theoretical value, itself measured on Jupiter and Saturn .... However, this excess of CO and HCN is not present on Uranus. Hence a double question: what is the origin of this excess on Neptune, and why is it different on Uranus? There is no definitive answer as yet to these questions.

Here is another unexpected discovery: observations conducted in 1997 by the European satellite ISO (Infrared Space Observatory) discovered the unexpected presence of oxygen constituents(water and carbon dioxide) in the stratosphere of the four giant planets and Titan. Where could they come from? At the tropopause, the temperature is low enough, on the four giant planets, for the water to be frozen. The tropopause acts therefore as a cold trap, which the atmospheric gases coming from the interior cannot get around. If water is present in the stratosphere of giant planets, it must come from outside. What is the origin of this flow of oxygen? Two sources are possible: a local source, from the rings and satellites of the planet, and an interplanetary source in the form of a stream of cometary-type, water-rich meteorites. The debate is not settled, but currently it appears that both sources must be considered. In particular, the collision of the comet

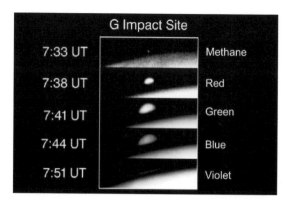

Figure 5.13. Observation of the collision of a fragment of Comet Shoemaker–Levy 9 with Jupiter in July 1994. The matter was raised up to an altitude of 3000 km. The image was taken by the Hubble Space Telescope (© NASA).

Shoemaker-Levy 9 with Jupiter, in July 1994, has illustrated that this type of event exists and probably contributes to the phenomenon we observe. The European Herschel satellite, in orbit since May 2009, has allowed us to study in more detail the stratospheric water in giant planets to better monitor possible temporal variations and constrain its origin.

5.1.5 An internal structure which is still unknown

At pressures above a dozen bars, the interior of the giant planets is not accessible to remote sensing. Regarding *in situ* measurements, the only information we have comes from the Galileo probe, which explored the deeper layers of the Jovian atmosphere down to a pressure of 22 bars. We have only indirect measurements of the giant planets' physical parameters: mass, radius, density, and flattening components of the gravitational field. From these data, theoretical models can be constructed by taking into account the changes of state of the material at high pressure; determining equations of state at high temperature and high pressure require complex laboratory experiments.

In the case of Jupiter and Saturn, the models suggest a three-layer structure: an envelope of molecular hydrogen, an ocean of metallic hydrogen and a central nucleus, very dense but fluid, made of heavier elements. At the center of Jupiter and Saturn, the pressures reach 40 and 10 million bars, and

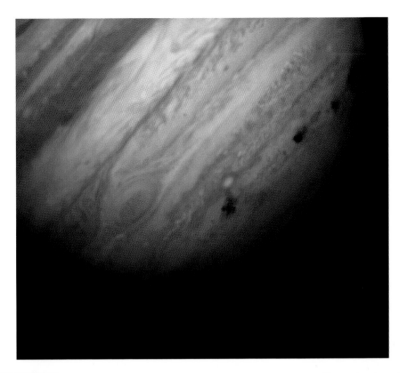

Figure 5.14. A few impact sites impact following the collision of the comet Shoemaker–Levy 9 with Jupiter (July 1994). As the twenty cometary fragments penetrated the atmosphere, they formed impacts along a parallel of latitude 44°S. The impacts remained visible for several months. The image was taken by the Hubble Space Telescope (© NASA).

the temperatures there would be 20,000 and 10,000 K respectively. The magnetic field of the two planets could be generated in the fluid core or in the ocean of metallic hydrogen.

Uranus and Neptune are a slightly different case: according to the equations of state of hydrogen at high pressure, the pressure conditions inside the two planets would not be sufficient for the hydrogen to be in liquid form. According to a theoretical model that remains hypothetical, their internal structure would be made of an envelope of molecular hydrogen, a fluid mixture of hydrogen and ice, and finally a core of rock. The central pressure is of the order of 8 million bars and the temperature is about 8000 K. The magnetic field could be generated in the mixture of hydrogen and ice or in the fluid core.

Finally, let us mention a special characteristic of three giant planets, Jupiter, Saturn and Neptune: the existence of an internal energy source. The Voyager probes have shown evidence for it, and only Uranus has no detectable internal source. In the case of the other three giant planets, the emitted energy is about twice the energy received from the Sun. The most plausible origin is the release of the energy stored during the formation phase (during which the planets were much larger and warmer than today). The energy is released through a slow contraction as the planet is cooling. In the case of Jupiter and Saturn, another contribution could come from the condensation of helium in metallic hydrogen: this condensation is accompanied by a precipitation of helium inward (and therefore a relative helium depletion in higher layers) and an outward release of energy.

The following paradox remains to be explained: while Uranus and Neptune seem so similar (as we have seen above, they may even have formed at the same distance from the Sun), why is Uranus lacking an internal energy source? For this difference, as for other singularities we will discuss later, we have no definitive answer yet.

5.1.6 A diversity of magnetospheres

We have seen (Section 3.6) that the Earth has a magnetic field generated by the dynamo effect within its liquid core. The resulting magnetosphere acts as a shield that protects the Earth's atmosphere from the solar wind, in the form of an hemispherical cavity facing the Sun and a long tail in the opposite direction.

The giant planets also have a magnetosphere whose main structures are similar to that of the Earth. We have seen above (Section 5.1.5) that the planets have a central part which is fluid and conductive: in the case of Jupiter and Saturn, it may be made of metallic hydrogen; for Uranus and Neptune whose internal pressures are lower, it could be an ocean of molecular hydrogen and other ionized molecules. In the case of the four planets, the dynamo must be amplified by their rapid rotation around their axis (between 10 and 16 hours).

Among the four planets, Jupiter's magnetosphere is the closest to the Earth. Its existence was discovered in the 1950s by the detection of a very

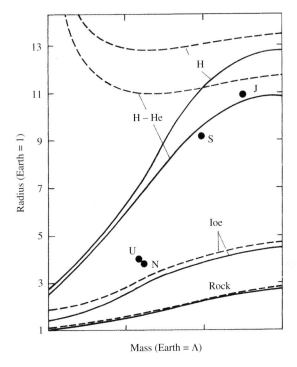

Figure 5.15. The mass–radius relation for a self-gravitating object, for different compositions. Jupiter and Saturn are in the field of the gas giants, while Uranus and Neptune are mostly composed of ice (after Stevenson, 1982).

intense radio radiation at 22 MHz, characteristic of synchrotron radiation of electrons in the magnetic field of the planet. Later, measurements of the space probes Pioneer 10 and 11, Voyager and Ulysses, Galileo and finally Cassini helped to refine its structure. As in the case of the Earth, the magnetic field of Jupiter is dipolar, with an axis slightly inclined with respect to the axis of planetary rotation. This dipole undergoes disturbances related to the presence of the Galilean satellites, in particular the closest one, Io. Like the terrestrial one, Jupiter's magnetosphere shows Van Allen belts, which are the source of synchrotron radiation. In polar regions, the interaction of energetic particles from the solar wind with the magnetosphere of Jupiter produces aurorae: atoms and ions in the ionosphere excited by energetic particles from the solar wind emit, in the visible spectral range, spectacular and colorful radiations.

Like Jupiter, Saturn has a magnetic dipole that has been revealed by space probes. The dipole axis is nearly aligned with the axis of planetary rotation. Saturn's magnetosphere differs from that of Jupiter by the presence of Titan and its hydrogen torus, and also by the presence of the ring system which prevents the trapping of particles along the field lines; this explains the absence of Van Allen belts. Saturn exhibits auroral phenomena similar to those of Jupiter. Finally, during its two successive flybys of the planets in 1986 and 1989, the probe Voyager 2 revealed the special nature of the magnetospheres of Uranus and Neptune. The dipoles are indeed very eccentric and inclined relative to the poles of the planets, indicating a more complex structure with multipolar terms. In the case of Uranus, the situation is further complicated by the exceptional position of the axis of rotation of the planet, almost in the plane of the ecliptic. Each magnetosphere is unique and presents a precious laboratory for plasma physicists.

5.2 From Jupiter to Neptune: four decades of exploration

Space exploration of giant planets began with two U.S. probes Pioneer 10 and 11 launched in the 1970s. Using simple instrumentation, they sent us the first spectacular images of the Great Red Spot of Jupiter and the rings of Saturn. A few years later, the highly successful Voyager mission gave us the foundation of our current understanding of the giant planets. Two identical probes, Voyager 1 and 2, were launched in 1977. Voyager 1 flew by Jupiter in 1979 and Saturn in 1980, while Voyager 2 flew by the four giant planets in 1979, 1981, 1986 and 1989. The exploration of Jupiter continued with the Galileo mission, launched in 1989, with an orbiter and a descent probe which entered Jupiter's atmosphere in 1995, providing the first *in situ* measurements of the thermal and cloud structure of the planet. The Galileo orbiter remained in operation around the Jupiter system until 2003. The Saturn system has been explored by the Cassini mission, jointly developed by NASA and ESA, and launched in 1997. It came with an orbiter, still in operation, and a European descent probe, Huygens, which landed successfully on the ground of Titan on January 14, 2005. Other projects are in preparation for the future exploration of Jupiter and Saturn,

Figure 5.16. The Great Red Spot of Jupiter, photographed by Voyager 1 in 1979. The origin of the orange color is still poorly understood. The white spot is a cloud of ammonia. (© NASA).

but nothing is currently planned for Uranus and Neptune (see Section 1.3.2). Note that ground-based observations, particularly in the infrared and millimeter range, have provided us with valuable information on the atmospheric composition of the giant planets. This is also the case for Earth-orbiting satellites, with IUE (International Ultraviolet Explorer), HST (Hubble Space Telescope), ISO (Infrared Space Observatory) and more recently Spitzer and Herschel.

5.2.1 Jupiter, the giant

Jupiter, the most massive of the giant planets and also the closest, has always been a prime target for telescopic observation. Cassini, from the late seventeenth century, oversaw its Great Red Spot and its circulation system into bands and zones generated by the rapid rotation of the planet. As on other planets, the difference in exposure between the equator and the poles generates a Hadley circulation (see Section 3.4), characterized by an alternating structure in bands and zones, which become more numerous as the planet's rotation period increases. The pale yellow color of the bands is due to the presence of ammonia clouds. The origin of the red color of the Great Spot is not clear; it could be due to the presence of sulfur or phosphorus constituents, or aerosols from methane photochemistry. The archives of the Paris Observatory mention the observation by Cassini of a series of impacts, which most likely came from the fall of a comet. A similar event, followed by the entire astronomical community, took place in July 1994: comet Shoemaker-Levy 9, in orbit around the planet, had broken into twenty fragments, as a result of tidal forces at perijove prior to its passage two years earlier. The twenty fragments then traveled in orbits very close together, to successively plunge into the atmosphere of Jupiter during the next perijove passage. The impacts have led to spectacular phenomena: local elevation of temperature above 10,000 degrees; ejection of matter over 3000 km altitude, dark craters observable for several months, and formation of new molecules (H_2O, CO, HCN, CS, OCS ...) by shock chemistry. The size of the initial core was estimated at about 1 km. The observation of this unusual phenomenon has allowed astronomers to study in real time (and safely!) the effects of a major meteoritic impact on a planetary atmosphere.

As mentioned above, Jupiter is the only giant planet to have benefited from *in situ* observations. Mass spectrometers and gas chromatographs were used to measure, at each altitude, the temperature, pressure, chemical composition, cloud composition and wind speed, as the probe descended into the atmosphere. Signals were sent down to a depth of 22 bars, and provided a unique database, which is still used as a reference today; it was also an unprecedented technological success. One of the most important results was the measurement of the abundance of the heavy elements with respect to hydrogen. For most of the gases, the enrichment was found to be about four times greater than the protosolar value, in full agreement with the predictions, thus validating unambiguously the model of giant planet formation from an ice core (see Section 5.1.2). Another spectacular result was the discovery of a very complex convective structure. The Galileo descent probe had entered in a very atypical subsidence region, particularly dry and devoid of clouds, a bit similar to tropical regions on

Earth. On Jupiter, these regions are also gathered along parallel strips as a result of the Hadley circulation. The mechanisms of this complex circulation are still poorly understood.

Although very weak, Jupiter, like the other giant planets, has a ring system, located less than two Jovian radii from the center of the planet. Accidentally discovered by Voyager 1 in 1979, and later observed again by the Galileo orbiter, the system consists of three components, made of very small particles of micron size; the rings are probably fed by the small satellites close to Jupiter.

In the wake of the Galileo mission, NASA has designed a new mission, JUNO, devoted to the measurement of the chemical composition in the interior of the planet, in order to constrain its formation scenario further. In addition, ESA has recently selected the very ambitious JUICE (JUpiter and ICy moons Explorer) mission, which will be devoted to the exploration of the Jupiter system and more specifically its moons Europa and Ganymede (see Sections 5.3.1 and 7.2.3). This mission is expected to be launched in 2022 and to approach Jupiter in 2030.

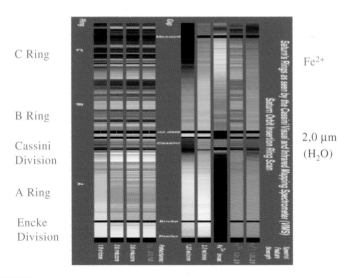

Figure 5.17. The composition of Saturn's rings. Although composed mostly of water ice, Saturn's rings also contain other elements such as iron oxides. According to Brown *et al.*, Astron. Astrophys., 2005.

5.2.2 Saturn, the most beautiful one

Saturn, like Jupiter is a gas giant: the protosolar gas constitutes about 90% of its total mass. It therefore has strong similarities with Jupiter in terms of thermal and cloud structure and meridional circulation associated with belts and zones. There is one difference, however: the lower temperature leads to the condensation of ammonia in larger quantities than in the case of Jupiter, which probably explains the more uniform pale yellow color of the planet.

The imposing ring system of Saturn makes it one of the most beautiful objects in the sky. Its variable appearance, depending on the position of the Earth relative to the equatorial plane of the planet, continues to fascinate amateur and professional astronomers, several centuries after Christiaan Huygens understood its nature. As telescopic observations were made more precise, new rings and new divisions were discovered. The thickness of the rings is very small: less than one hundred meters. Voyager 1, in 1980, revealed for the first time the extreme complexity of the rings of Saturn. Images have shown the presence of thousands of individual rings constantly evolving, with different eccentricities, interacting with each other; their evolution is linked to that of neighboring satellites, which play a role of confinement, hence their name of "shepherd satellites." Rings, such as satellites, are mainly composed of water ice, with particles of all sizes ranging from micrometers to meters. The lifetime of individual rings and small neighboring satellites is very short, and may not exceed a few months or a few years; tidal forces are responsible for these changes.

Following the Cassini mission, another project has been studied by NASA and ESA. The TSSM (Titan Saturn System Mission) mission was designed for the exploration of Saturn, its ring system, and also Titan and Enceladus, two fascinating satellites for astronomers (see below, Sections 5.3.2 and 7.2.5). At ESA, this ambitious mission was not selected, as it came as a second priority to the JUICE mission, but its concept could emerge again in a couple of decades.

5.2.3 Uranus and Neptune, the fraternal twins

The two icy giants, Uranus and Neptune, are similar in size and density. Their colors (dark blue for Neptune, pale green to Uranus) are probably

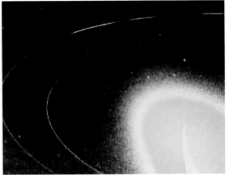

related to the presence of methane in abundance, but the cause of the difference in color between the two planets is not clear. At first glance, Uranus and Neptune are twins; yet, when observed more closely, several questions arise. Firstly, their orbital position is unexpected. Why is Neptune, more massive than Uranus, it is located at a greater distance from the Sun? We have seen that a possible explanation of this paradox could be the migration of the giant planets early in their history; however, this hypothesis remains to be confirmed (see Section 6.1.3).

There are between two planets several unexpected differences. The first concerns their internal energy. Uranus has none, while the other three giant planets emit an energy that is about twice the solar flux they receive (see Section 5.1.5). Why are Uranus and Neptune different? Here is another anomaly: Neptune's stratosphere contains abnormally high amounts of CO and HCN, which is not the case of Uranus, nor Jupiter and Saturn. What is its origin? The question is open. Another difference: Neptune exhibits strong dynamic activity that induces a very active weather and intense vertical motions. This is also the case for Saturn: the two planets have a relatively warm lower stratosphere and strong turbulent activity. Finally, we have to mention the exceptional nature of the geometry of Uranus, whose axis of rotation is nearly in the plane of the ecliptic. This peculiarity could be the signature of a collision that occurred at the beginning of history of the planet, before the formation of regular satellites in the equatorial plane of the planet; alternatively, according to numerical simulations, it might also result from a chaotic dynamical evolution of the system of Uranus.

What are the possible causes of differences between Uranus and Neptune? It has been suggested that the internal convection of Uranus could be inhibited, which would have the effect of preventing the removal of the internal energy contained in the deeper layers, and remove any dynamic activity. This would explain the absence of internal energy, turbulence and meteorological activity. Still, the cause of inhibition remains to be understood.

The rings of Uranus and Neptune, both very weak, were discovered only recently (1977 and 1984 respectively). In both cases, they were observed from Earth during a stellar occultation experiment, during the passage of each planet in front of a star (see Section 1.2.2). In the case of Uranus, before and after the transit, the light curve of the star showed a series of nine very short occultations, signatures of the existence of nine very tenuous rings. At the time of the observation, these were seen in a nearly circular form, because the axis of rotation of the planet was oriented in the direction of the Sun and the Earth. Voyager 2, in 1986, confirmed the presence of these rings. In the case of Neptune, the light curve of the star was more surprising: occultations were detected, but only on one side. The astronomers then interpreted the observations as the presence of arcs. In 1989, Voyager 2 showed that the rings did exist throughout the orbit, but they showed an increased thickness in some places, as a result of their interaction with small neighboring satellites. Like those of Saturn, the rings of Uranus and Neptune are composed of water ice, but they are darker. They are probably covered by a deposit of organic material resulting from the irradiation of icy grains of several kinds by energetic particles from the magnetosphere.

5.3 The outer satellites

Unlike the terrestrial planets, the giant planets are surrounded by numerous satellites. These fall into two broad categories. The regular satellites, very close to the equatorial plane of the planet, were formed during the collapse

Figure 5.19. Io, the Galilean satellite closest to Jupiter, photographed by Voyager in 1989. This image shows the current satellite volcanic activity. (© NASA).

of the protosolar material around the original core (see Section 2.5); the irregular satellites, with random eccentricities and inclinations, were captured by the planet. Because of their strong gravitational field, the capture mechanism is much more efficient than in the case of terrestrial planets.

The outer satellites cover all scales of sizes, from a few kilometers to the size of Mercury for the biggest of them (Ganymede, Titan and Callisto). There is no clear boundary between the smaller inner satellites of Saturn and the larger fragments of the rings that are in constant interaction with one another (see Section 5.2.2).

The four giant planets have a great variety in the size distribution of their satellites. Jupiter has four large satellites — called Galilean since their discovery by Galileo in 1610 — at distances ranging from 6 to 26 Jovian radii. Uranus' satellites have a comparable configuration, with five large satellites located between 5 and 25 Uranian radii from the planet.

Saturn, however, has a single large satellite, Titan, located at 20 Saturnian radii, as well as eight smaller satellites with radii between 100 and 800 km. All the satellites mentioned above are regular. Neptune, in contrast, is a special case. It has a large irregular satellite, Triton, located at about 15 Neptunian radii, with a very high angle (160°) Versus Neptune's equator. It is probably a transneptunian object whose capture has had the effect of ejecting the previous regular satellite system. The list of outer satellites is continuously growing, due advances in ground-based telescopic instrumentation, and also, in the case of Saturn, thanks to observations from the Cassini orbiter, still in operation.

The outer satellites, becoming better known thanks to space exploration, are extremely diverse as a result of the nature, history and environment of each specific body: mass, density, distance to the planet, interaction with neighboring bodies... In addition to their own interesting properties, they can be seen as possible examples of what could be exosatellites or low-mass exoplanets far from their star, in other planetary systems. We will examine below some of them in more detail.

5.3.1 Close to Jupiter, the Galilean satellites

The four Galilean satellites (Io, Europa, Ganymede and Callisto, in ascending order of their distance from Jupiter) are small in size, but their lower density (less than $2g/cm^3$ for Ganymede and Callisto, about 3 g/cm^3 for Europa). This implies the presence of water in significant quantities, as can be expected for objects located beyond the snow line (see Section 2.5).

As an exception, Io, located at 6 Jovian radii from Jupiter, has a density of 3.5 g/cm^3, and its surface properties are very different from the other three Galilean satellites. This singularity is due to the strong tidal effects related to the gravitational field of Jupiter, over ten thousand times more massive than the satellite. These tidal effects, coupled with the gravitational effects of the nearby satellites Europa and Ganymede, in resonance with Io, are responsible for the active volcanism highlighted in 1979 by Voyager 1. Io's surface is constantly renewed, as shown by the absence of impact craters on its surface. Pictures taken later by the Galileo and Cassini missions, as well as by the HST, showed that volcanic activity on Io is evolving rapidly, on the scale of a few years. As evidenced by its high density, Io is not primarily composed of water, because water has escaped due to the temperature rise induced by the internal energy released from tidal effects. Io's surface is covered with ice sulfur dioxide; its atmosphere, fed by volcanism, is also composed of SO_2. Although very thin (a few nanobars at the surface), Io's atmosphere plays an important role in the

Figure 5.20.   Methane lakes of Titan as revealed by the Cassini radar (© NASA).

magnetic environment of Jupiter because it ejects ionized atoms of hydrogen, sodium, sulfur, oxygen that feeds a torus around the orbit of Io.

Europa, the smallest of the Galilean satellites is, like Ganymede and Callisto, covered with water ice. However it is also, to a lesser extent than Io, subject to the tidal forces of Jupiter. This results in an internal energy that has the effect of increasing the temperature sufficiently, according to models, for water under the surface to be in a liquid state down to the silicate core. Observations by Voyager and Galileo have confirmed this hypothesis. The spacecraft have observed, on the surface of Europa, series of cracks and plates arranged in a way suggesting that they are moving over a viscous or liquid medium. In addition, the Galileo orbiter magnetometer discovered a magnetic field, induced by the Jovian magnetic field, which could indicate the presence of a conductive medium, such as a saltwater ocean. Europa is thus an object of peculiar interest for exobiologists in the search for environments which might favor the emergence of life (see Section 7.2.3). Finally, Ganymede and Callisto, farthest from Jupiter and more cratered, have the characteristic density of external objects rich in water. The project JUICE mentioned above, recently selected by ESA, includes the exploration of the three icy satellites as its main scientific objective.

5.3.2 Titan, an analogue of the early Earth?

Close to the two outer Galilean satellites by its size and density, Titan, the largest satellite of Saturn, is nevertheless very different: it is the only satellite in the solar system with a substantial neutral atmosphere. Titan's surface is permanently hidden by a thick layer of orange–yellow aerosols, now attributed to condensation of hydrocarbons and nitriles. We knew very little about the atmosphere and surface of Titan prior to its exploration by Voyager 1 in 1980. This probe revealed that the atmosphere consists mainly of molecular nitrogen, and its surface pressure is 1.5 times the atmospheric pressure on Earth. These two properties were sufficient to stimulate the interest of exobiologists who see Titan as a possible laboratory for prebiotic chemistry, or even a possible analog of the early Earth (see Section 7.2.5). There is one major difference, however: the very low temperature prevailing on Titan (94 K at the surface, 70 K at the tropopause).

After nitrogen, methane, with a partial pressure of 2%, is the most abundant atmospheric constituent. The simultaneous presence of CH_4 and N_2 is responsible for a complex chemistry, different from those of the planets, and unique in the solar system. As in the giant planets, the photo-dissociation of methane leads to the presence of many hydrocarbons; in addition, the dissociation of N_2 by solar UV radiation and energetic particles from Saturn's magnetosphere leads to the formation of nitriles (HCN, C_2N_2, HC_3N, CH_3CN). Finally, Titan receives, like the giant planets, a stream of oxygen particles of ice from neighboring satellites and/or an interplanetary oxygen flux (see Section 5.1.4). All these ingredients combine together to create a complex chemistry.

Nearly twenty years after the success of Voyager 1, NASA and ESA jointly launched in 1997 the Cassini–Huygens mission, designed to explore the Saturn system. In July 2004, the Cassini spacecraft began orbiting Saturn. On 14 January 2005, the European probe Huygens transmitted to Earth the first images of Titan: a flat surface covered with a dark deposit of hydrocarbons and heavily eroded boulders, most likely composed of water ice. There was no trace of liquid methane, contrary to some predictions. However, Huygens, images showed valley networks suggesting the presence of past river systems. More recently, observations by the orbiter's radar have discovered the presence of lakes, mainly located at high northern latitudes; others have been also found at high southern latitudes. These lakes change over the seasons. When the northern hemisphere is in winter, lakes filled with liquid methane (and possibly other hydrocarbons) are more numerous, especially near the north pole. What is the source of methane in Titan? As it is photodissociated in the atmosphere, it must be constantly renewed. It has to be injected into the atmosphere from the sub-surface by cryo-volcanism (this mechanism consists of the ejection of a gas that condenses immediately due to the low temperature of the medium).

Another spectacular result of the Cassini mission is the evidence for an extremely complex chemistry in Titan's ionosphere. This result has been found by mass spectrometry, revealing dozens of complex ions exceeding one hundred units of atomic mass. This discovery led to a new challenge for astronomers who are now trying to reproduce in the laboratory the conditions of formation of these ions, and also to understand the

Figure 5.21. Saturn's satellite Enceladus, observed by the camera of the Cassini orbiter mission. We see the emission of a plume, rich in water vapor, ejected from the surface by cryo-volcanism. (© NASA).

exact nature of the condensates that form the haze obscuring the surface. Indeed, Titan keeps surprising us... hence the efforts of the planetary community in preparing a new exploration mission towards Titan, which could emerge in a couple of decades (see Section 7.2.5).

In view of its size and density, Titan is very close to Ganymede and Callisto. The three satellites are distant enough from their planet for its effect on their composition and internal structure to be minor (which is not the case for Io and Europa, closer to Jupiter). Why does Titan have a dense atmosphere while the Galilean satellites do not? The answer can probably be found in the differences in their heliocentric distances. Located just beyond the snow line, 5 AU away from the Sun, Ganymede and Callisto were mainly made of water ice: as mentioned above, water is the first molecule that condenses when the heliocentric distance increases (see

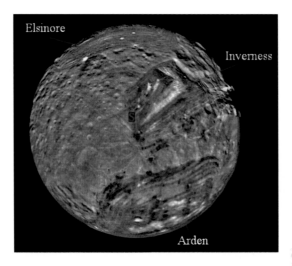

Figure 5.22. Uranus' satellite Miranda, photographed by Voyager 2 in 1986 (© NASA).

Section 2.5). Located at 10 AU from the Sun, Titan was formed at lower temperatures and could trap other ices, starting with ammonia (NH_3) and methane (CH_4). The ammonia quickly turns into molecular nitrogen N_2 under the effect of solar ultraviolet radiation, while the reverse reaction is very difficult. It is therefore very likely that the nitrogen of the atmosphere of Titan comes from the photodissociation of NH_3. Its molecular weight, higher than that of water, has favored the stability of the atmosphere.

5.3.3 From Saturn to Uranus

Among the five largest regular satellites ranging between Saturn and Titan (in order of increasing distance: Mimas, Enceladus, Tethys, Dione and Rhea), Enceladus deserves special attention. The Cassini magnetometer first detected a local atmosphere, outgassed from the south pole of the satellite. The camera of the Cassini orbiter then observed plumes of water vapor and organic compounds ejected from this place. Infrared observations then showed a local elevation of temperature in this region, also ribbed and younger than the rest of the globe. Enceladus is subject to cryo-volcanism activity, not fully explained yet (see Section 7.2.4). Recent models suggest the presence of pockets of liquid water near the surface.

Enceladus is another interesting object for exobiologists, and will be closely studied by the next space mission to Saturn.

The five large satellites around Uranus (by ascending distance to the planet: Miranda, Ariel, Umbriel, Titania and Oberon) are also covered with water ice but darker than in the case of Saturn's satellites, probably because of the presence of other ices (methane, hydrocarbons, etc.). During the Voyager 2 flyby in 1986, the first images of Miranda created a big surprise: this small satellite has a surface with highly rugged cliffs, faults, and valleys showing evidence for intense tectonic activity in the past. What may have been the source of internal energy in such a small and cold object? A possible hypothesis is that the satellite could had been destroyed in a collision, and then have reaccreted from fragments orbiting Uranus, but it is only a hypothesis.

5.3.4 Triton, another Pluto?

Finally, Triton, the largest moon of Neptune and the last sample of our gallery, deserves special mention. First, unlike all the objects mentioned above, it is irregular, as evidenced by its retrograde and highly inclined orbit. This peculiar geometry causes the strongest seasonal effects existing in the solar system. As seen from the surface, the Sun is at zenith at latitudes up to 52° north and south! As in the previous cases, the orbit of Triton is synchronous: it always presents the same face to Neptune. The nature of Triton was revealed by Voyager 2 during its 1989 flyby. The atmosphere is, like that of Titan, composed of nitrogen with a small component of methane; however, the surface pressure is only a few microbars. With a temperature of 38 K, the surface of Triton is the coldest and the brightest one observed in the solar system. Triton's surface is young and shows signs of a cryo-volcanism which may still be active today. Infrared spectroscopy measurements made from Earth revealed the presence of ices: N_2, CH_4, CO, CO_2 and marginally H_2O.

We have seen that Triton is likely a transneptunian object captured by Neptune. The conditions of this capture are not completely clear to dynamicists. However, as a possible proxy of Pluto, Triton is of special interest to us, as it has been closely observed by Voyager 2. The New

Horizons spacecraft, launched in 2006 by NASA, will reach Pluto in 2015.

Observations of Pluto and its satellite Charon, made from the ground since the 1980s, have only confirmed the remarkable similarity between Triton and Pluto. Several techniques have been successfully used: observations of mutual occultations by Pluto and Charon, stellar occultations (see Section 1.2.2), infrared spectroscopy, millimeter measurements and observations in Earth orbit, with the HST and ISO. Between 1979 and 1998, Pluto, whose orbit is significantly more eccentric than the giant planets', was closer to the Sun than Neptune and Triton. The (very relative) elevation of temperature resulted in the sublimation of the ices, particularly nitrogen and methane. A gaseous atmosphere was then detected on Pluto, very similar to that of Triton, also consisting of N_2 and a small contribution of CH_4, with a surface pressure of a few microbars. On the surface of Pluto, the ice composition is also very similar to that of Triton. That reinforces the hypothesis that Triton is also a transneptunian object. It remains to understand how Triton was trapped in the orbit of Neptune: a problem still open for dynamicists.

6

Exoplanets, the New Worlds

Are we alone in the Universe? This question has been explored by humanity since the dawn of civilization. It divided the Greeks between supporters, as Epicurus, of an infinite number of inhabited worlds, and those believing in the uniqueness of the Earth, the center of the universe, as claimed by Aristotle. The debate was then purely philosophical but already agitated minds.

With the advent of the Copernican system, in the sixteenth century, the question was reconsidered in a scientific context. The observation of the stars showed that the Earth and planets revolve around the Sun, which is itself a star like billions of others. Therefore, how can we not imagine that some of these stars could also be surrounded by a planetary system? In 1600, for having supported this hypothesis, too bold for its time, Giordano Bruno, a victim of religious dominant dogmatism, paid for his conviction with his life. But the idea continued to make its way. In 1686, the philosopher Fontenelle, in his "Conversations on the Plurality of Worlds" strongly advocated not only an infinite number of planets in the Universe, but also the habitability of some of them, including in our solar system. Still in the seventeenth century, the astronomer Christiaan Huygens, in his book "Cosmotheoros", also argued for the existence of alien planets. A century later, the philosopher Immanuel Kant and the physicist Pierre-Simon Laplace lay the groundwork for what will become the "model of the primitive nebula". According to this theory, the Sun and planets of the solar system were formed within a disk resulting from the collapse of a rotating interstellar cloud (see Chapter 2). If this scenario is valid for the solar system, why would not it be the case for other stars? We find this assumption in the nineteenth century in the "Popular Astronomy" of the astronomer Camille Flammarion, a true reference book of the time.

6.1 A long quest marked with failures

If planets exist around nearby stars, how can we detect them? Very soon, astronomers realized the extreme difficulty of the task. Let us take the case of Jupiter and see if we could observe it if we were located at the distance of the nearest neighboring star. This star is Proxima Centauri, located at 1.3 parsec (*), i.e. about 4 light-years away (this means that it takes 4 years for its light to reach us). The angular separation between the Sun and Jupiter, as seen from this distance, is 4 arc seconds, which would be easily measurable. However, what is almost impossible is to resolve the light from the planet, which is drowned in the solar light. In the visible range, the Sun is a billion times brighter than Jupiter! As the global temperature (130 K in the case of Jupiter) is much lower than that of the Sun (5600 K), the flux of Jupiter increases in the infrared, and the contrast between the solar flux and planetary becomes more favorable. Still, at a wavelength of 10 microns, the Sun remains a million times brighter than the planet. Today, techniques have been developed to try to extract very weak signals near bright sources, as we will see later (see Section 6.5.1). But in the nineteenth and twentieth centuries, such observations were impossible. Astronomers of that time therefore turned to methods of indirect detection.

6.1.1 Early research by astrometry

If a star is surrounded by a planet of sufficient mass, its motion is affected: the star–planet system rotates around its center of gravity with a period which is the revolution period of the planet around the star. The more massive the planet, the more affected the stellar motion; the closer

---

*The parsec is the distance at which one astronomical unit is seen on the sky with an angle of 1 arc second.

the planet, the shorter the orbital period. Measuring the stellar motion requires excellent accuracy and great patience: it consists in identifying very precisely the position of the star which might have a planet, as compared with other fixed neighboring stars. The presence of the planet can then be identified through a small elliptical motion of the star on the celestial sphere.

This method, called astrometry, had already allowed the astronomer Friedrich Bessel, in the nineteenth century, to detect the presence of a low-mass stellar companion around the star Sirius. Armed with this success, some astronomers decided to apply the same technique to search for exoplanets. The most famous astronomer was Piet Van de Kamp, who announced in 1944 the discovery of the first exoplanet located around Barnard's Star. It turned out later that the exoplanet did not exist: the motions of the star, misinterpreted as due to the presence of a planet, actually resulted from an instrumental problem related to the telescope used for the measurements. In 1974, P. Van de Kamp announced a new detection, this time around the star Epsilon Eridani, but again this result was not confirmed. It now appears that the instrumental means of the time did not provide sufficient astrometric accuracy given the difficulty of the challenge.

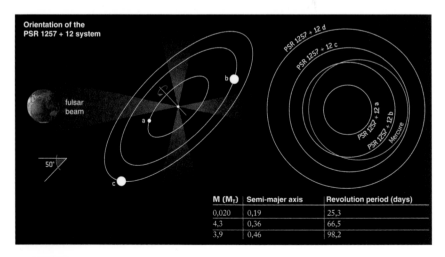

M (M_T)	Semi-majer axis	Revolution period (days)
0,020	0,19	25,3
4,3	0,36	66,5
3,9	0,46	98,2

Figure 6.1. The planetary system around the pulsar PSR 1257 +12 (© F. Casoli and Encrenaz T., Extrasolar planets, Belin, 2005).

6.1.2 Around pulsars, strange planets

By the early 1990s most astronomers had abandoned the search for exoplanets by astrometry, but the question remained a burning issue. Another method, completely different, showed up: it consists in studying the perturbations around the periodic radio signal emitted by pulsars. What are they? Pulsars ("pulsating stars") are neutron stars, very small and extremely dense; residues of a supernova explosion, they have reached the final stage of their existence. They have a very strong magnetic field, and were detected in 1967 through the radio signal they emit, which is extremely stable in frequency.

By the 1970s, radio astronomers were interested in the possible disturbances of this periodic signal, as they might be possible signatures of the presence of a companion. Several attempts were announced, but no detection could be confirmed. Finally, in 1992, the astronomer Alexander Wolszczan announced the discovery of two planets around the pulsar PSR 1257+12, whose rotation period is very short (1.3 ms). A third planet was subsequently detected. Two planets, with orbits close to that of Mercury, have masses close to that of the Earth, while the third, closer to the star, is a hundred times less massive.

1992 is thus the year of the discovery of the first exoplanetary system: the discovery was a big surprise. Multiple questions were raised: what is this strange star with a planetary system? It has nothing in common with the Sun. Pulsars with millisecond periods (known as "millisecond") are very old stars, unlike the first pulsars which were discovered with a period close to one second, and which are much younger. Pulsars "millisecond" could be the last stage of the evolution of a system composed of a normal star and a neutron star. As the older star dilates, it gradually transmits its material to the neutron star whose rotational speed accelerates. What do such objects look like? It is very difficult to describe them; at most, we can advance that they do not resemble the solar system, and in terms of habitability, they are not the candidates we are looking for.

6.1.3 The advent of velocimetry: First discoveries and first surprises

Despite this success, astronomers remained unsatisfied. Will we find some day exoplanets resembling the planets we know? To proceed in the

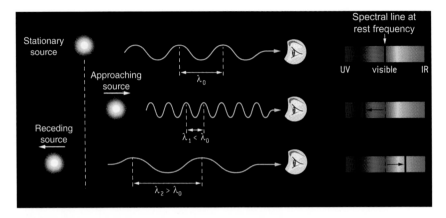

Figure 6.2. The principle of the Doppler effect. For a source moving toward the observer (middle), the wavelength decreases (i.e. the frequency increases). If the source recedes (bottom), the opposite effect takes place (© F. Casoli and Encrenaz T., Extrasolar planets, Belin, 2005).

direction, it was first necessary to focus on solar-type stars. We also had to develop a new technique, more sensitive than astrometry. This new method, velocimetry, came to maturity in the early 1990s.

As astrometry, the velocimetry technique consists in detecting the motion of the star with respect to the center of gravity of the star–planet system. But instead of measuring the stellar motion on the celestial sphere, astronomers determine the velocity of the star, as measured from Earth, relative to the center of gravity. To do so, they use the Doppler effect. Its principle is the following: if a star is approaching the observer, the frequency of the radiation it emits is shifted towards shorter wavelengths, (i.e. high frequencies); if the star moves away, its radiation is shifted towards longer wavelengths (or low frequencies); this is the famous redshift, observed in distant galaxies, which has shown evidence for the expansion of the Universe. Applied to our star–planet system, the method consists in measuring in the visible range, with a very high spectral resolution, the Doppler effect of a number of spectral signatures of the star (we use well-identified lines of its spectrum, whose wavelengths are very precisely measured). The observation is repeated over time in the hope of seeing a periodic curve.

Astrometry and velocimetry are complementary methods to measure the motion of the star relative to its center of gravity. Astrometry measures

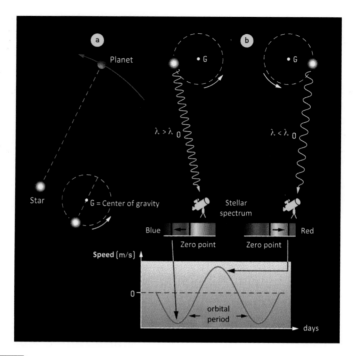

Figure 6.3. The principle of velocimetry. The motion of the star is considered with respect to the center of gravity of the star+planet system (© F. Casoli and Encrenaz T., Extrasolar planets, Belin, 2005).

the motion on the celestial sphere, whereas velocimetry measures motion in the perpendicular direction, along the line of sight of the observer. If a planetary system has its axis of rotation aligned with the line of sight, velocimetry detects no motion; in contrast, astrometry sees a circular displacement. If the axis of rotation is located on the celestial sphere and perpendicular to the line of sight, velocimetry records a maximum deviation of the speed, while astrometry measures only a slight motion of the star along a segment, and not an ellipse. Incidentally this shows a limitation of the velocimetry method: as it only measures the velocity component aligned along the line of sight, the inclination to the planetary system as seen by the observer is not known. The method does not actually determine the mass of the exoplanet, but only a lower limit of it.

As in the case of astrometry, the velocimetry method requires patience, since the observations must cover the entire period of revolution of the putative planet, or at least a significant fraction of it. Searching for the equivalent of Jupiter is not an easy task: let us remember that its revolution period around the Sun is nearly 12 years! Searching for the equivalent of the Earth requires a year of observation, but the expected speed modulation is much lower, since the planet is less massive. For example, the presence of Jupiter on the Sun induces a solar velocity modulation of 12.5 m/s, whereas that induced by the presence of the Earth is only 10 cm/s.

In the early 1990s, high-resolution spectrographs, combined with telescopes larger than 1 meter in diameter, allowed us to measure, for stars near the Sun, typical speeds of the order of a few m/s. Detecting an "exo-Jupiter", if it exists, is in principle feasible. But will we have to wait ten years for a complete revolution of the planet around its star?

The answer is no, fortunately. Here comes an unexpected discovery: there are many giant exoplanets in close proximity to their stars: their period of revolution is only a few days! Such objects are much easier to detect than "exo-Jupiter", since their detection can be suspected within a few weeks, and confirmed within a few months. Thus, in 1995, came the big event: the first discovery of an exoplanet around a solar-type star. The monitoring program was carried out at the Observatoire de Haute

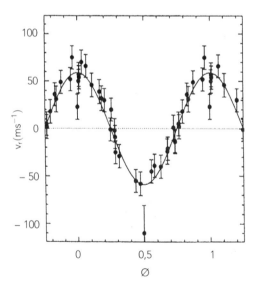

Figure 6.4. Velocimetry curve of the star 51 Peg; the oscillation is due to the presence of a planet (after M. Mayor and D. Queloz, Nature, 1995).

Provence by two astronomers from the Geneva Observatory, Michel Mayor and Didier Quéloz. The new exoplanet was found orbiting the star 51 Pegasi (and was thus called 51 Peg b) with a mass equal to at least half that of Jupiter, and a revolution period around its star of 4 days: its distance from the star is 0.05 AU! The discovery had a considerable impact in the astronomical community and beyond. In the following weeks, two other exoplanets were discovered by Geoffrey Marcy's team in the United States: they are located around 47 Uma and 70 Vir, have masses at least equal to 2.6 and 7 times that of Jupiter, and their average distances to their star are 0.5 AU and 2 respectively. In the coming months, discoveries of new exoplanets will multiply; in most cases, they are giant exoplanets very close to their host stars.

The extreme interest raised by these findings can be explained by two reasons. The first one is beyond the scope of astronomy: at last, an answer is given to the question explored by mankind for centuries: there are many planets around solar-type stars, and the potential implications for the

Figure 6.5. The La Silla Observatory in Chile. The 3.60m telescope is shown at the top. It hosts the HARPS spectrometer, the most accurate velocimetry instrument to date (© ESO).

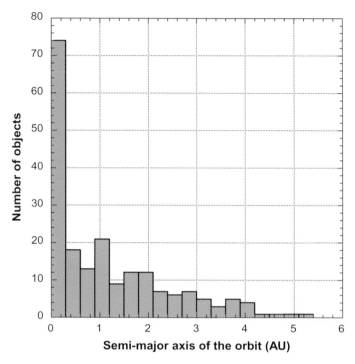

Figure 6.6. Histogram of the first extrasolar planets according to their distance to the star (© M. Ollivier).

search for extraterrestrial life are huge. But beyond this discovery, the first question that comes is: what is the nature of these exoplanets? It was expected that, as in the case of the solar system, giant planets would be found at large distances from their stars, since they presumably formed at temperatures sufficiently low to allow the accretion of an ice core. But the giant exoplanets are very close to their stars. Imagine Jupiter turning around the Sun with a period of four days... its distance from the Sun would be 10 solar radii, and its maximum angular distance would be 5 degrees; its angular diameter would be more than 3 arc minutes, ten times less than the Sun.

From the early discoveries, a conclusion appears: new planetary systems are very different from ours. It remains to understand why, and to better characterize these exotic objects. This is the challenge of current and future research.

6.1.4 Another successful technique: planetary transits

In the few years following the detection of 51 Peg b, velocimetry was the only method for the detection of exoplanets. Another technique, also indirect, emerged in 1999: it consists of observing the host star as the exoplanet passes in front of it; the event is called a transit. It requires a peculiar geometrical configuration: the Earth must be located in the plane of the planetary orbit that is then observed edge-on. Of course, the planet is too small to be detected by itself, but its presence in front of the stellar disk results in a slight decrease in the stellar flux. If observed from outside the solar system, Jupiter transiting in front of the Sun would induce a decrease of 1% in the solar brightness, since the diameter of Jupiter is approximately one-tenth of the Sun's. In the case of the Earth, ten times smaller, the decrease would be only 0.01%.

It is still necessary that the planet pass in front of its star ... What is the probability of such an event? It increases with the angular size of the star, and is therefore much higher if the star has a large diameter or is close to the Sun. Transits are also more frequent if the planet is close to its star, which is precisely the case of a large number of recently discovered exoplanets. Thus, the probability of observing an exoplanet like 51 Peg b (also sometimes called the "Pegasids" or "hot Jupiters") is about 10%; however, in order to observe a planet similar to Earth, the probability is only 0.5%. The duration

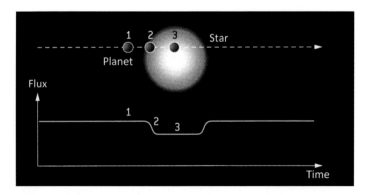

Figure 6.7. Diagram of a planetary transit. When a planet passes in front of the host star, the stellar flux decreases by a small fraction (about 1% for a Jupiter-type planet) (© F. Casoli and Encrenaz T., ibid.

of the transit is on the order of fifteen hours for an exoplanet located at 1 AU from its star, and about twice as much if it is located at 5 AU.

The transit method consists in measuring continuously, with a high stability, the radiation from stars of a given stellar field over a long period, a few weeks at least. In principle, the detection of giant exoplanets by transit is possible from Earth, because the stability of ground-based photometry measurements allow us to detect variations at the percent level. However, the detection of exo-Earths is not possible from the ground, and requires observations from space. In the case of giant exoplanets, ground-based measurements still suffer from the difficulty of continuously observing a given stellar field, unless the observation is made from high latitudes in the direction of one of the poles.

In comparison to the velocimetry technique, transit measurements have an extra advantage: the exoplanet's diameter can be inferred. The velocimetry technique usually gives a lower limit of the exoplanet's mass, because the indination angle of the planetary system is unknown. In the case of a transit, this angle is known, so the velocimetry technique actually gives the exact mass of the exoplanet with the diameter measurement retrieved from the transit observation, the density can then be inferred. This provides us with constraints about the nature of the exoplanet (gas, ice or rock).

In 1999, a US team led by David Charbonneau announced the detection of the first exoplanet in transit. The observation was subsequently

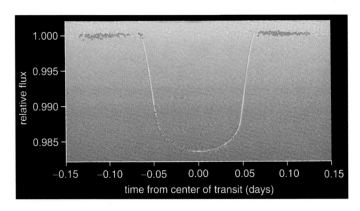

Figure 6.8. Light curve of HD209458b, the first discovered transiting exoplanet, observed with the HST (after Brown *et al.*, 2001).

confirmed by measurements made by the Hubble Space Telescope. The exoplanet was given the less-than-poetic name of HD209458b. It is a hot Jupiter, whose mass is two-thirds that of Jupiter, with a very low density (0.3 g/cm$^{3)}$. Following this success, many observational efforts from the ground were initiated. They led to the detection of more than over one hundred exoplanets by transit.

How to find the exo-Earths? We have seen that the required photometric precision is beyond the capabilities of ground-based observations. This is why some space mission projects have emerged. The first is the French CoRoT mission, launched in December 2006, which has been operating in Earth orbit over a period of six years. CCD cameras are recording stellar fields, without interruption and with great stability, over a period of a few weeks to a few months. About 25 exoplanets have been detected, and many potential candidates are waiting for confirmation

Figure 6.9. The CoRoT satellite, launched by CNES in December 2006, dedicated to the search for transiting exoplanets (© CNES).

Chapter 6. Exoplanets, the New Worlds

using velocimetry. The CoRoT mission ended in November 2012. A U.S. satellite with greater sensitivity, the Kepler mission, was launched in 2009 and operated until May 2013, with the objective of detecting planetary transits. Four years after its launch, the Kepler mission has been a tremendous success. Over two thousand candidate exoplanets have been discovered; most of them are too faint for confirmation by velocimetry, but they can be used for statistical studies; the number of confirmed exoplanets is close to one hundred. The TESS mission, to be launched by NASA in 2016, and the CHEOPS mission, selected by ESA for a launch in 2017, will be the next steps of this research.

Figure 6.10. The Kepler mission, launched by NASA in March 2009. Kepler has firmly detected about a hundred transiting exoplanets, and more than two thousand candidates have been identified (© NASA).

6.1.5 A promising emerging technique: direct imaging

For a long time, direct imaging of exoplanets has been beyond the capabilities of available observational means, for the reasons mentioned above: the flux contrast between the planet and the star is too faint. Still, it is possible to enhance this contrast by adding together several favorable factors. The detection is made easier if a faint star is chosen, if the exoplanet is massive and far from its star, and if the observation is made in the infrared range, where the planet/star flux ratio is enhanced. These favorable factors were all combined for the first detection of an exoplanet by direct imaging. The exoplanet 2MASS 1207b, detected by G. Chauvin and his team at Grenoble Observatory, was observed in 2005 by direct imaging at the Very Large Telescope of ESO with the adaptive optics system NAOS-CONICA and a coronagraphic system (see Section 6.5.1). The object, whose mass is 5 times that of Jupiter, is located at 55 AU from its star. It is a young object of a few million years, and therefore hot. The star is a brown dwarf, with a mass equal to 0.025 solar masses. It is relatively cold, which made the detection possible: the contrast between the stellar and planetary fluxes is particularly favorable, and the distance between the exoplanet and the star corresponds to 778 ms of arc, measurable with current techniques. Since the detection of this exoplanet, about thirty objects have been detected by direct imaging. This technique is expected to flourish in the coming decade with, in particular, the operation of the SPHERE instrument on the VLT, whose first light is expected at the end of 2013.

6.1.6 Other detection techniques

In parallel with velocimetry and transit observations, another original method has also led to the detection of several exoplanets. It is based on an effect of general relativity, called "gravitational lensing": when a star passes in front of a distant star, the light rays coming from the distant source are curved by the mass of the star and converge towards the observer to produce a lens effect. This effect, observationally demonstrated nearly a century ago, was used, in particular, to try to detect the missing "dark matter" in our galaxy by finding brown dwarfs. This research has not proved fruitful; however, the same effect, renamed "microlensing", can be used to detect the presence of an exoplanet around the star, used as a lens in front of a distant object. In this case, the amplification curve of the flux of the distant object has on its wing a characteristic peak corresponding to the passage of the exoplanet surrounding the lens. Several international efforts have led to the detection of microlenses and a dozen exoplanets have been detected. The advantage of this method is its high sensitivity; we can identify a few Earth masses of exoplanets only. The disadvantage of this method is that the event is not repetitive nor predictable, as it depends on the fortuitous passage of the star in front of a distant object.

To summarize this overview of the methods presently used for detecting exoplanets, let us try to to identify their relative merits:

- Astrometry is particularly suited for the detection of massive objects away from their stars. A major step in this area is expected with the astrometric space mission Gaia, to be launched in 2013 by the European Space Agency.

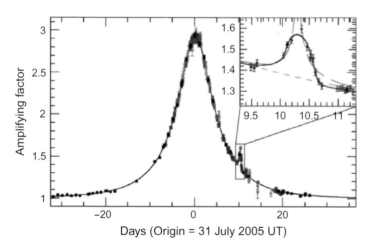

Figure 6.11. The light curve of a star detected by microlensing. When a planet is present, an oscillation is visible in the wing of the amplification curve. The figure shows the detection of a small and cool planet around a M-dwarf. Adapted from Beaulieu *et al. Nature* **439**, 437, 2006.

- Velocimetry is the preferred tool for detecting massive objects close to their star, as illustrated by the spectacular results obtained during the last fifteen years. The improvement in high-resolution spectrographs have allowed us to decrease the mass limit of detected exoplanets down to only a couple of Earth masses, and performances continue to improve.
- The timing of pulsars has a very high sensitivity relative to the mass of the detected exoplanets. The method is however limited to a very special class (pulsars with planets) which has very few objects. The total number of planets detected around pulsars is less than ten.
- The observation of planetary transits has increased sensitivity for large objects close to their star, and is limited to a particular geometrical configuration. When combined with velocimetry, it allows the determination of the physical parameters of the exoplanet (mass, radius, density).
- The technique of gravitational microlensing is very sensitive with respect to the mass of the detected objects; in particular, it is suitable for large objects more than one AU away from their stars. It allows the detection of exoplanets around distant stars located in the halo of our Galaxy. The physical parameters of the exoplanet are determined from model fitting. The limitation is the lack of repeatability of the event.
- Finally, the direct imaging method is especially well adapted for massive planets located far from small and faint stars (i.e. M-dwarfs). The method has led to the detection of about 30 objects and will further develop in the future with the operation of new instruments designed for this purpose.

Other methods are being studied to detect exoplanets directly: coronagraphy, interferometry, and the search for radio signals of exoplanets. They are described in more detail below (see Chapter 6.5).

6.2 Fifteen years later, the situation

We are now more than fifteen years after the discovery of the first exoplanets. Today (mid 2013), we know than more about 900 exoplanets. Most of them were discovered by velocimetry, but an increasing number is being found by transit. Meanwhile, the detection of super-Earths is increasing; there are currently over a hundred. It is now possible to derive statistical properties for these objects: their mass, distance to the star, and radius (hence density) in the case of exoplanets in transit, as well as the properties of their host stars and those of the planetary systems themselves. What is the proportion of stars with planets? Does it depend on their spectral type, their mass, their content in heavy elements? What is the probability for a star to have a multiple planetary system? These are the questions that we will discuss below, keeping in mind the observational biases that favor the discovery of massive exoplanets close to their star.

6.2.1 Giant exoplanets close to their stars!

The first discoveries have already shown it: the proportion of detected giant exoplanets close to their star is surprisingly high. Nearly half of all detected exoplanets are located less than 0.4 AU from their star and, among these, about a third orbit at 0.05 AU or less, which corresponds to a rotation period of 4 days or less. This is the case, in particular, for the "hot Jupiters." We will come back later to this new class of objects, unknown in the solar system, and the implications for models of stellar and planetary formation (see Section 6.3.3). The proximity of hot Jupiters to their star, which implies a synchronous rotation with them (i.e. always presenting the same side to the star) has also serious consequences for the planets themselves, as it implies strong contrasts between the day and night sides.

6.2.2 A majority of exoJupiters?

The velocimetry method allows the detection of exoplanets, but also, more generally, of all low-mass companions orbiting their stars. We have already mentioned (see Introduction) that these companions can be classified according to their mass. If it is less than 0.01 solar masses (or 13 Jupiter masses) it is called a planet and its internal energy is not sufficient to generate the first cycle of thermonuclear reactions leading to stellar nucleosynthesis. If the mass of the object is between 0.01 and 0.08 solar masses (between 13 and 80 Jupiter masses), it is called a "brown dwarf": the core temperature is sufficient to initiate the first cycle of deuterium destruction, but not the

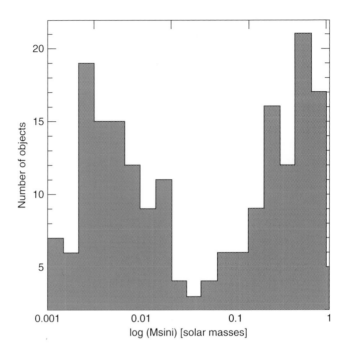

Figure 6.12. The mass distribution of stellar and substellar companions. The left part of the curve corresponds to exoplanets, the right part to small stars. The gap in the middle of the curve is known as the brown dwarf desert; it illustrates that planets and stars are formed following very different mechanisms (after Santos *et al.*, *Messenger*, **110**, 32, 2002, © ESO).

next one which transforms hydrogen into helium. Beyond 0.08 solar mass, the object is a star, in which a complete cycle can be set up (even faster if the initial energy is important). The stars are classified into various types (O, B, A, F, G, K, M), as a decreasing function of their mass and temperature; their lifetime is shorter as the temperature is higher. Our Sun, a G-type star, is in the middle of the range; its lifetime is about ten billion years. Being about 5 billion years old, the Sun is at the middle of its life.

If we trace the mass distribution of stellar companions detected by velocimetry, we see a clear bimodal distribution with a first mass peak between 0.001 and 0.01 solar mass (corresponding to the exoplanets) and the other beyond 0.1 solar masses (the classical stars). The absence of objects between the two peaks, interpreted as the "brown dwarf desert", highlights the very low number of objects belonging to this category.

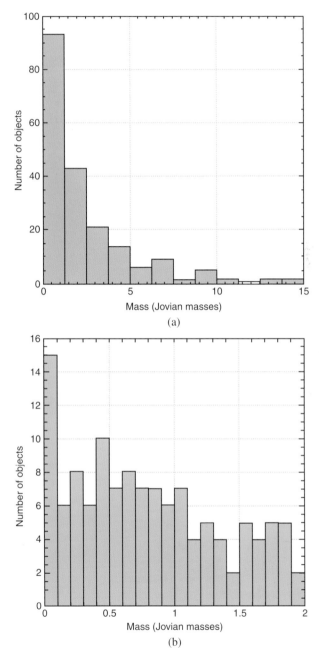

Figure 6.13. Distribution of exoplanets based on their mass: (a) $M < 15\,M_J$; (b) $M < 2\,M_J$ (© M. Ollivier; www.exoplanets.eu).

Let us go back to the histogram of the exoplanets' mass. While velocimetry *a priori* favors the detection of the most massive objects, we observe that the most numerous ones have the mass of Jupiter: there are very few items with more than two times the mass of Jupiter. Note that, in order to reach this conclusion, we must make an assumption about the average inclination of the systems that we observe; the assumption is that this inclination takes all possible values without any preferred angle. Between 0.1 and 2 Jupiter masses, the distribution is evenly distributed. As for the number of items of mass less than one tenth of the mass of Jupiter, the number is increasing rapidly with the progress of velocimetry and transit techniques: they are the "exoNeptunes".

How does the exoplanets' mass depend upon the distance to their star? Two conclusions emerge: firstly, there are few supermassive objects close to their star (in other words, the "hot Jupiters" are not "super-Jupiters"; in addition, there is a large number of exoplanets with a mass comparable to Jupiter between 0.5 and 5 AU. All these observational facts should be taken into account in the construction of formation scenarios of planetary systems.

6.2.3 Very eccentric exoplanets

With the solar system, we have become used to planetary orbits that are nearly circular; This is no more the case in the realm of exoplanets, with the exception of objects very close to their star, "hot Jupiters" with periods shorter than 6 days. The circular orbit of the latter may be the result of tidal effects due to the strong gravitational field of their stars. Another consequence is probably the synchronous rotation of these objects, already mentioned above.

For objects with periods longer than a few days, no correlation appears between the eccentricity and the distance to the star. Some eccentricities have very high values, exceeding 0.90. A possible explanation is to be found in the interactions between the exoplanet and the protoplanetary disk, which become predominant with respect to tidal effects, as we will discuss below (see Section 6.3.3). Finally, note that we do not see any correlation between the eccentricity of exoplanets and their mass; this is also a constraint for formation models.

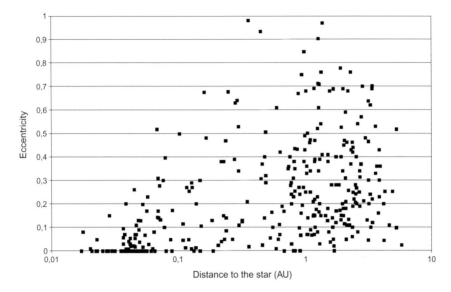

Figure 6.14. Eccentricity distribution of exoplanets, as a function of their distances to the star. In the vicinity of the host star, the orbits tend to become circular due to tidal effects (© M. Ollivier; www.exoplanets.eu).

6.2.4 Multiple planetary systems

Several tens of planetary systems have been detected, with up to 7 components. Currently, about 12% of the observed planetary systems have at least two planets. Given the observational difficulty of identifying such systems, it is possible that multiple planetary systems are at least as common as simple systems, and even more. The solar system, in this respect, is not an exception.

Orbital characteristics of components of multiple systems can, like other exoplanets, include high eccentricities. Several systems have resonances: the periods of revolution of the planets in the system are in a ratio of integers. This stable configuration could be the result of migration: we have seen how, in the early history of the solar system, a moderate migration of the giant planets led the Jupiter–Saturn system into a 2:1 resonance, which had important implications for the dynamics of all small bodies of the solar system (see Section 5.1.3).

In multiple systems, resonances may cause very important dynamical effects. Two planets located in resonance encounter growing gravitational interactions, which may eventually lead to the ejection of one of the two objects. Multiple systems resonance are especially interesting for specialists of gravitational dynamics who study their orbital evolution.

6.2.5 The role of heavy elements

If the search for exoplanets was first focused on solar-type stars, it quickly spread to a larger variety of spectral types. We now know of planetary systems around not only stars of types F, G, K and M, but also around young objects, multiple stars, not to mention pulsars. We can already see that the ability to form a star surrounded with a planetary system is not a unique characteristic of solar-type stars.

It is possible to study the probability of a star having an exoplanet based on the fraction of heavy elements it contains, which is quantified by a parameter called the metallicity. What is it? We know that according to cosmic abundances, hydrogen is the most abundant element by mass; the mass fraction X, which represents it, is approximately 0.75. The helium mass fraction, called Y, is about 0.23. All other elements called "heavy elements" constitute the remaining 2%: this is called Z, the rate of metallicity. We have seen that the heavy elements are produced in stars, then they return into the interstellar medium at the end of the star cycle, following the explosion of novae or supernovae; the heaviest elements are synthesized during supernova explosions. Metallicity varies from one star to another. The cosmic abundances are determined from solar abundances, the only ones that can be accurately measured. For other stars, their metallicity is measured from the signatures of metal atoms (e.g. iron) present in their visible or ultraviolet spectra.

What can we learn by studying the number of exoplanets around stars of different metallicity? The probability that they have a giant exoplanet is much higher if they are rich in metals. This conclusion, *a priori*, is not very surprising: in the solar system, the planets are formed by accumulation around solid cores. If heavy elements are more abundant, there is more solid mass available to form large nuclei, which in turn can capture the surrounding gas. However, the correlation between the metallicity of

Chapter 6. Exoplanets, the New Worlds

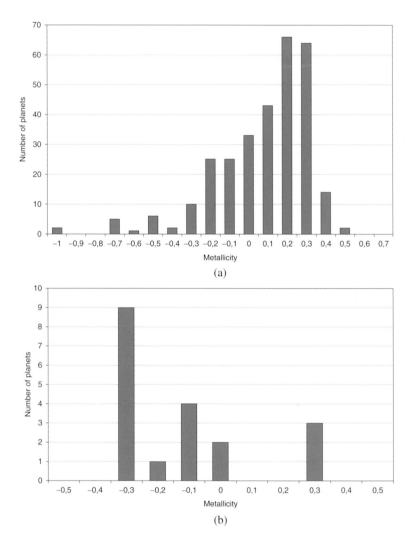

Figure 6.15. Distribution of exoplanets as a function of their metallicities: (a) Planets with masses greater than 10 Earth masses (b) Planets with a mass below 10 Earth masses (© M. Ollivier; www.exoplanets.eu).

the star and the number of planets is not clear in the case of super-Earths; this will require a larger sample to be specified.

There is another unexpected result: as the metallicity of the Sun is relatively low, this is *a priori* not favorable for the formation of a planetary system. Maybe we can explain this paradox if we consider that massive exoplanets are more easily detectable. The correlation we observe therefore rather concerns giant exoplanets, which would form more easily around large nuclei in the case of a star with high metallicity. We still know very little about Earth-like exoplanets.

6.2.6 How many stars with planets?

In the light of all the information collected over the past fifteen years, we can now provide an answer to this simple question: what is the probability for a star to have a planet? Beyond the astronomy community, it is indeed a matter of strong concern for the public, since its answer affects the probability of the existence of extraterrestrial life.

Here are the results that we have today, based on current observations, their interpretation and extrapolation. Of all the stars of type F, G and K observed, 7% have at least one giant exoplanet with 5 AU. Following the most recent statistics from Kepler, more than 50% of stars have an exo-Earth or a more massive exoplanet. If we consider now the stars whose metallicity is twice that of the Sun, the probability of detecting an exoplanet is 25%.

Of course, these results are very preliminary. One must take into account the observational bias in favor of massive stars of short period, and the specificity of the various selected samples. We now have to refine the study, get access to lower masses, and expand the study to other spectral types of stars and other specific sampling criteria. We are still at the beginning of a fascinating research field, which will focus particularly on low-mass exoplanets — the famous "super-Earths".

6.2.7 Searching for super-Earths

At the beginning of exoplanet exploration by velocimetry, the performance of high-resolution spectrometers limited this technique to the search for giant exoplanets. Over the past recent years, it has become clear that the advancements in the sensitivity of these instruments now make possible the detection of smaller objects, of less than ten Earth masses. This is a new field of research for ground-based astronomy, which has already been successful.

While the limit of detectability in the radial velocity was about 10 m/s at the time of detection of 51 Peg b, we can now achieve an accuracy of 50cm/s, which allows us to detect objects whose mass is only a couple of Earth masses. This involves choosing particularly stable stars, devoid of intrinsic fluctuations, as they provide a disturbance that limits the detectability of stellar companions. Thus, during the last three years, the team of Michel Mayor, with the HARPS spectrometer installed at the 3.60 m telescope at ESO in Chile, has detected several exoplanets of very low mass. The exo-Earths are within our reach!

Advances in the velocimetry technique are especially valuable as they allow us to monitor transiting low-mass exoplanets detected by transit. We have seen that the combination of methods allows a complete characterization of the objects: mass, diameter, density, distance to the star. This coupling will be fruitful in the coming years with the analysis of the CoRoT and Kepler results. In the near future, direct imaging techniques, currently under development, will also become operational.

6.2.8 A small sample of exotic objects

Among the 850 exoplanets known today, astronomers have searched with special care for those who might have similarities with ours. So far, they have not really found any. This can be explained, as we have seen, by the observational bias that favors massive objects close to their star. But along the way, they discovered some really weird objects. Here are a few examples, arbitrarily chosen among many other exotic objects. As new discoveries are reported every month, if not every week, more and more exciting discoveries are expected to surprise us in the near future.

6.2.8.1 *HR 209458b*

Here is HD 209458 b, the first discovered transiting exoplanet, already mentioned above: this object is the perfect example of the "hot Jupiter" (note that the name Osiris, used at the time of its detection, is no longer found in the literature today). The planet has a mass equal to 0.7 times that of Jupiter, and a radius equal to 1.3 times the Jovian radius, corresponding to a very low density of 0.4 g/cm^3. For example, the density of Saturn, the least dense planet in the solar system, is almost twice that value. However the planet's originality mainly lies in its short orbital period (3.5 days), placing it at a distance of 0.045 AU from its star. So close to its star, the planet is most likely in synchronous rotation, and therefore always presents the same side to the star. The temperature difference between day and night sides could be very strong, and generate winds of several thousand km/h. Subject to such intense stellar radiation, can its atmosphere be stable? Nothing is less certain. Theoretical models predict a continuous evaporation of gas that could form a long tail in the opposite direction to the star, as in the case of a comet. If this is the case, all of the gas could disappear in a few billion years, which could also be the fate of many other hot Jupiters.

In an attempt to analyze the atmospheric composition of the object, spectroscopic measurements of the star were made during and outside transits. From the observation of primary transits (when the planet passes in front of the star), hydrogen was detected (which is not a surprise given the low density) but also oxygen, carbon, and sodium that could indicate the presence of a deeper cloud of sodium. From 2005, secondary transits have been also observed (corresponding to the passage of the planet behind the star). The Spitzer satellite measured the temperature of the exoplanet's dayside (more than 1000 K) and detected water vapor, while the HST measured its spectrum in the near infrared and discovered water vapor and methane. From the point of view of its chemical composition and its atmospheric dynamics, HD 209458b is a rare pearl for observers as well as theorists.

6.2.8.2 *Corot-7b*

With a radius of 1.7 times that of Earth and a mass of 4.8 Earth masses, the seventh exoplanet detected by the CoRoT satellite has a density of 4.7 g/cm^3, not very different from the Earth. But the similarity stops here: CoRoT-7b is extremely close to its star, at 0.017 AU, and its orbital period is less than a day. Like HD 209458 b, Corot-7b always presents the same face to the Sun and must show extreme day/night contrasts. With a temperature above 1000 K, the dayside may be made of a magma rock, while the night side, at a temperature of about 100 K or less, would be made of rock and possibly ice. Note that velocimetry measurements have allowed the detection of a second super-Earth orbiting the CoRoT-7 star: located at 0.046 AU from the star, CoRoT-7c has a mass of 8.4 Earth mass. As for CoRoT-7b, it is one of the most extraordinary objects discovered to date.

6.2.8.3 *OGLE-05–390L: the success of microlensing*

We remain in the field of small exoplanets. Now here is one of the first objects detected by the microlensing technique. It has a mass of 5.5 Earth masses and is located at 2.6 AU from its star. Observed in 2006, it kept for over two years the record of the smallest mass (if we exclude planets around pulsars). The modeling of the light curve of the distant object allows us to determine the orbital and physical parameters of the exoplanet (see Section 6.1.4). It is unfortunately difficult to say more, because the phenomenon of microlensing cannot be repeated on the same object.

6.2.8.4 *The system HD 69830: the three Neptunes*

We now turn to multiple systems. The HD 69830 system, detected by Lovis in 2006, has the unique property of hosting three planets with masses between 5 and 20 Earth masses. The orbits are located at 0.08, 0.2 and 0.6 AU. In addition, a dust disk located at less than 1 AU must be present, as shown by the infrared flux excess measured in the spectrum of the star. Dynamical models show that such a system can be stable over a period of a billion years. What may be the nature of the three planets? The innermost could be rocky, while the outermost one may have a gaseous envelope surrounding a core of ice and rock. Given the distance to the star, water might possibly exist in liquid form on the surface of such an object. This system is thus of special interest for astrobiologists.

6.2.8.5 *KOI-55b and KOI-55c: Two very small dense cores*

These two planets, detected in 2011, are among the smallest exoplanets detected so far. They form a compact system very close to the post-red giant hot star KOI-55, located at distances of 0.0060 AU and 0.0076 AU respectively from the star, with revolution periods of 5.76 and 8.23 hours respectively. The star is a pulsating B subdwarf, in an evolved stage of helium nuclear fusion, which shows a complex pulsating spectrum, well analyzed by the Kepler satellite. The presence of the two planets was inferred from an in-depth analysis of the Kepler light curve, showing evidence for extra planetary fluxes probably due to the reflected component

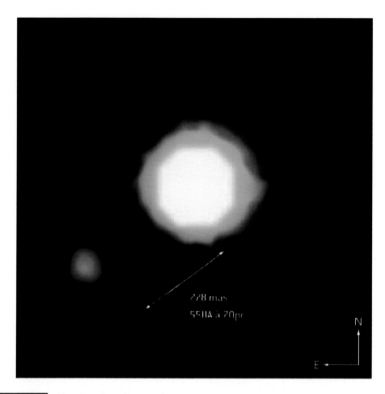

Figure 6.16. First detection of an exoplanet by direct imaging. It is a companion of 5 Jovian masses, located 55 AU away from the dwarf 2M1207, whose mass is equal to 0.025 solar mass. The flux contrast between the planet and the star is particularly favorable. The observation was made with the NACO instrument on the VLT (G. Chauvin *et al.* 2005) (© ESO).

of the stellar light. KO1-55c, the smallest planet known to date, has a mass of about 0.7 terrestrial masses and a radius of 0.85 terrestrial radii. KOI-55cb is more massive and much denser, with a mass of 4 terrestrial masses and a radius of about 0.7 terrestrial radii. The objects might be the leftover dense cores of former giant planets engulfed in the red giant envelope in the late stage of its evolution.

6.2.8.6 *GJ 1214b: A temperate super-Earth or a mini-Neptune?*

Discovered in 2009, the transiting exoplanet GJ 1214b has intriguing properties. Located at 0.014 AU, the object orbits a M-star with a period of 1.6 days. Due to the low effective temperature of the star (about 3000 K), the equilibrium temperature of the planet (about 500–600 K) makes it a temperate object. With a mass of 1.7 times the Earth mass, and a radius of 2.7 terrestrial radii, the exoplanet might be either a hydrogen-rich mini-Neptune or a rocky super-Earth with possibly a water-rich atmosphere. This exoplanet is thus an ideal target to be analyzed by spectroscopy during transit. Unfortunately, GJ 1214 is a faint star, of 14th visible magnitude, which makes such observations a challenge with present capability. Still, these observations will become feasible in the future (see below, Section 6.5). Atmospheric models and theoretical spectra have been developed, and there is abundant literature about this object.

6.2.8.7 *Planets around multiple systems*

Since 2010, a new class of strange objects has appeared: exoplanets around binary systems, and even planets in orbit around a multiple system component. Besides the difficulty of finding an adequate nomenclature for these objects, their existence raises intriguing questions with respect to the stability of these systems. Several systems have been found, with one or even two planets orbiting binary systems. An even stranger system has been discovered: a transiting circumbinary planet (PH1) in a quadruple stellar system (Kepler-64)... Regarding such exotic systems, we can only anticipate more and more surprises.

6.3 The formation of planetary systems

In the light of the observations accumulated over fifteen years, and given the diversity of the discoveries, are we able to define a consistent scenario of the formation of planetary systems? While many questions remain unanswered, it is nonetheless possible to define the outline of the formation of stars and planetary systems. The two go together because exoplanets are born in a protoplanetary disk which accompanies the formation of the star itself; this scenario also applies to the solar system.

6.3.1 The scenario of star formation

Stars are born continuously in our Galaxy and in others. Initially, a fragment of interstellar cloud, denser than its environment, can contract under the influence of its own weight, as a result of an instability (e.g. the explosion of a nearby star). This fragment, where the temperature does not exceed a few tens of Kelvin, is composed of gas and dust; the chemical elements are present, at first approximation, according to their cosmic abundances: hydrogen is dominant, then helium, and molecules such as H_2O, CH_4, NH_3, CO, HCN, H_2CO,..., as observed in the interstellar medium. The solid phase comprises silicates and carbon aggregates, possibly covered with ice.

As the cloud fragment contracts, its temperature and density increase, as well as its speed of rotation. If the mass of the cloud is sufficient, the increasing rotation leads to the collapse of the cloud into a disk, perpendicular to the rotation axis of the cloud. The center of the disk will form the protostar; this is also the scenario that we described for the solar system (see Chapter 2).

Our knowledge of the early stages of star formation is based on two kind of data: the observation of young stars, and that of the associated

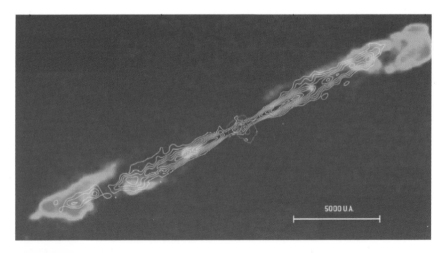

Figure 6.17. A protoplanetary disc and bipolar jets observed by mapping the CO molecule by millimeter interferometry; the young object HH is 211 observed here with the Plateau de Bure Interferometer at IRAM (© IRAM).

Chapter 6. Exoplanets, the New Worlds

protoplanetary disks. These disks are detected through their infrared and millimeter radiation; the CO molecule is used as a tracer of molecular abundances. Millimeter measurements, performed at very high spectral resolution, allow a very precise measurement of the gas velocities. They have shown the existence of very violent bipolar jets emitted along the axis of rotation of the disks. In parallel, the UV, visible and infrared observations of very young objects of T-Tauri type (according to the name of the star in which this behavior was first observed) shows the existence of mass ejection with huge mass loss rates among objects younger than ten million years. The mechanisms responsible for this mass loss can be stellar winds, magnetic fields or turbulent viscosity in the disk. Direct observation of such objects is difficult because they are often hidden in a cocoon of gas and dust; many of their envelopes have been observed by the Hubble Space Telescope.

6.3.2 Protoplanetary disks

The observation of young stars around us suggests that the formation of protoplanetary disks around forming stars is a common scenario in the Universe, not limited to our own solar system. Here is another remarkable result: these disks have a limited lifetime, of the order of tens millions years. It seems that the intense activity of young stars in their T-Tauri stage has the effect of ejecting outward most of the material. This implies a strong constraint on the model of planet formation in the disk, since they must have completed their growth before the disappearance of the surrounding gas.

How can a planet form in the protoplanetary disk? The outline of the scenario is similar to the one of the solar system formation (see Chapter 2). Initially, following the collapse of the cloud, the dust tends to migrate toward the plane of the protoplanetary disk, where the density increases sharply. Collisions are frequent, but can be non-destructive, because the relative velocities between particles are small: all particles rotate together in the plane around the protostar. Particles can clump together by electrostatic interaction. The growing process is probably fractal, as observed for interplanetary dust collected in the stratosphere by spacecraft. According to the simulations, it is possible to generate grains of several centimeters in a few thousand years.

What is the process allowing these grains to reach kilometer size? This problem remains poorly understood at present. Some models involve a gravitational instability in the disk, others favor the continuation of the clotting mechanism, and the turbulence of the gas may also play a role. The point is that, one way or another, the disk generates a multitude of kilometer-size "planetesimals". At this stage, gravitational interactions come into play: the largest of these fragments can attract the surrounding grains by gravity, sweeping the material around their orbit. The models show that this type of "runaway growth" promotes the formation of a small number of large objects in nearly circular orbits, in a few tens of thousands of years. Collisions between these objects will still decrease their number to the benefit of the largest ones.

How do giant exoplanets form in the disk? The question still divides the community. The most widely adopted scenario implies, as for the giant planets of the solar system, the accretion of matter around a solid core whose size is sufficient to capture the surrounding gas. For the core to be large enough (at least ten Earth masses), it must be made of ice, which implies a temperature lower than about 180 K. The accretion thus takes place at a safe distance from the star (more than 3 AU in the case of a solar-type star). This is where we find the paradox mentioned above: how to explain the presence of giant exoplanets close to their stars?

6.3.3 Migration within a disk

The answer to the paradox is probably migration. This process allows a giant planet to move either inward or outward by the effect of interactions with the gas disk, the dusty disk, or other planets. The effect is known in the solar system, where it has been invoked to explain the existence of resonance in outer satellite systems, or that of trans-Neptunian objects (the Plutinos) in resonance with Neptune. We have seen that in the case of the solar system, the migration of the giant planets has been limited. But *a priori* nothing prevents giant exoplanets, by this mechanism, to start from outside, beyond the snow line where it was formed by nucleation, to approach very close to its star.

Commonly cited models take into account the interaction between the planet and the gas disk. The interaction creates a torque between the disk and the planet that allows energy transfer resulting in some cases in a decrease of the distance of the planet, which becomes then closer to its star. For a planet whose mass is small compared to that of the disk, the decay time is estimated at about 100,000 years. In the case of a massive planet, a groove is created in the disk of gas around the orbit of the planet, and the decay time is even shorter: a few thousand years! In theory, the migration mechanism is perfectly capable of bringing the giant exoplanet close to its star. But a different question arises: how to stop the migration? If the migration times are so short with respect to the lifetime of the protoplanetary disk, all exoplanets should be swallowed by their star and we should not observe many of them. There is another troubling fact: a large number of hot Jupiters are located at 0.05 AU from their star. Is there a mechanism to stop the migration at this distance? Some dynamic models predict that after a phase of high eccentricity, the planet stabilizes at a steady distance equal to twice the Roche limit (which is 2.5 times the radius of the star), in a semi-circular orbit. Other processes may come into play: In the strong gravitational field of the host star, the planet may lose mass (as has been suggested in the case of exoplanet HD 209458b) or be distorted by tidal effects. Regarding the final stage of the exoplanets' migration, we can for the moment only speculate.

6.4 How to classify exoplanets?

What do we know about the exoplanets we have discovered? From velocimetry observations, we know their distance from their star and their revolution period, and we have a lower limit on their mass. For transiting objects, we also know the radius and density. But we know nothing of their physical nature. A few spectroscopic observations have allowed us, in some cases, to detect some atoms or molecules: hydrogen, oxygen, sodium, methane, water. But the information is still fragmented.

In the case of the solar system, thermochemical equilibrium models tell us about the nature of the molecules that we can expect to find under given conditions of temperature and pressure. Carbon and nitrogen are preferably found in the form of CO and N_2 at high temperature and low pressure, while CH_4 and NH_3 are expected under the opposite conditions. This is consistent with the fact that the atmospheres of the terrestrial planets consist mainly of CO_2 and N_2 (hydrogen is too light to be captured), while CH_4 and NH_3 are present in the atmospheres of the giant planets. Let us extrapolate these models to giant exoplanets: for a solar-type star, CH_4 and NH_3 should be predominant at a distance of a few AU; CO should become dominant over CH_4 below 0.1 AU, and N_2 would be more abundant than NH_3 below 0.05 AU. Water (H_2O) should be present also in the form of gas or ice, or even liquid if the temperature of the medium is above 0°C, in the appropriate range, depending on the pressure. At deeper levels, other elements may be present in the atmosphere in solid form (clouds, aerosols): silicates, titanium oxide, metals.... All these models assume that the elements are present in their cosmic abundances.

From our knowledge of the planets and satellites of the solar system, let us define some broad classes of objects that we could use to try to set up a classification of exoplanets. Of course, it is only an extrapolation, and we will wait for new measurements to confirm or refute these hypotheses.

Although solar system objects are extremely diverse, it is still possible to define some broad classes. Two parameters are essential: the mass of the object and its temperature (directly related to its heliocentric distance). Regarding the mass, we consider two types of objects: those with masses below ten Earth masses and those who are more massive (as the latter are able to capture the surrounding gas and form the so-called "giant planets").

In the category of small objects, we have the "rocky planets" and "icy planets" (these are the outer satellites). In both categories, the objects may or may not be surrounded by an atmosphere, which can be dense (in the case of Venus or Titan) or thin (in the case of Mars or Triton). What is the nature of the atmosphere, where it exists? In the case of rocky objects, it is dominated by CO_2 and N_2 (we leave aside the case of the Earth that has accumulated oxygen because of the emergence of life). In the case of icy objects, it is dominated by N_2 and CH_4 (this is what we see on Titan, Triton and Pluto). Why do we find molecular nitrogen N_2, and not NH_3? The reason is probably that ammonia NH_3 is readily converted to N_2 by photodissociation, while the reverse reaction is very difficult. Note that, in the case of icy planets, the atmosphere has a stratosphere, following the dissociation of N_2 and CH_4, which form the hydrocarbons and nitriles. This is not the case for rocky planets, except the Earth whose stratosphere is created by the presence of ozone, as a photodissociation product of oxygen.

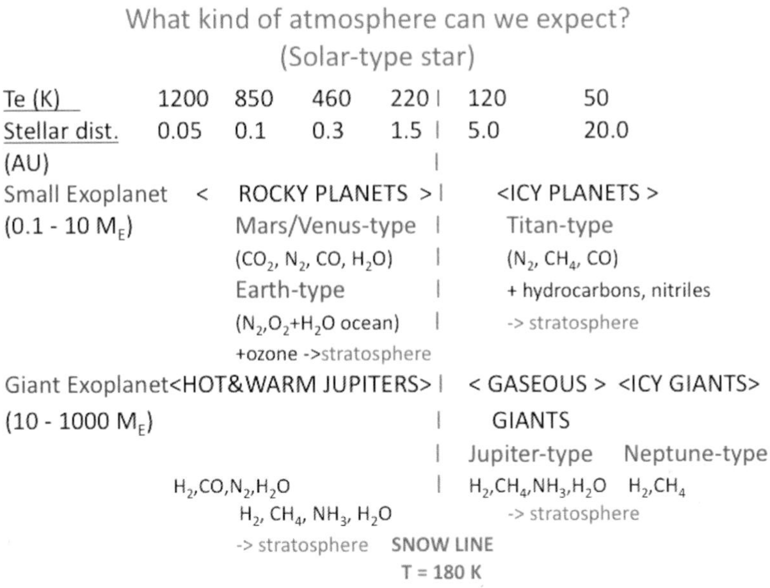

What kind of atmosphere can we expect?
(Solar-type star)

Te (K)	1200	850	460	220	120	50
Stellar dist. (AU)	0.05	0.1	0.3	1.5	5.0	20.0

```
Te (K)           1200  850    460    220 | 120        50
Stellar dist.    0.05  0.1    0.3    1.5 | 5.0        20.0
(AU)                                    |
Small Exoplanet  <  ROCKY PLANETS  > |    <ICY PLANETS >
(0.1 - 10 M_E)      Mars/Venus-type  |      Titan-type
                    (CO_2, N_2, CO, H_2O) |  (N_2, CH_4, CO)
                    Earth-type       |      + hydrocarbons, nitriles
                    (N_2,O_2+H_2O ocean) |   -> stratosphere
                   +ozone ->stratosphere
Giant Exoplanet<HOT&WARM JUPITERS> |   < GASEOUS >  <ICY GIANTS>
(10 - 1000 M_E)                    |     GIANTS
                                   |     Jupiter-type   Neptune-type
         H_2,CO,N_2,H_2O          |     H_2,CH_4,NH_3,H_2O   H_2,CH_4
           H_2, CH_4, NH_3, H_2O  |        -> stratosphere
           -> stratosphere   SNOW LINE
                             T = 180 K
```

Figure 6.18. Different types of atmospheres expected for exoplanets. From a few simple parameters (mass of the planet, distance to the star, spectral type of the star), it is possible, by comparison with the planets of the solar system and assuming thermochemical equilibrium, to make an assumption about the nature of their atmospheres.

Let us now turn to the category of massive planets. We can distinguish the gaseous giants and the icy giants, according to the relative proportion of gas and ice they contain. Their atmospheres, in all cases, are dominated by hydrogen and methane; they have a stratosphere rich in hydrocarbons, products of methane photodissociation.

Now let us try to project this simple classification into the realm of exoplanets. We initially assume a protoplanetary disk with a mass comparable to that of the protosolar disk, i.e. a low mass disk (about one hundredth of the mass of the star). Let us start with a solar type star. We assume that exoplanets have a average albedo of 0.3, a typical value encountered in the solar system. In the category of small planets, we have three possible cases: below 0.4 AU, a rocky planet with no atmosphere (unless very massive); between 0.5 to 3 AU, a rocky planet with an atmosphere dominated by CO_2 and N_2 (this category corresponds to the potential habitable planets); beyond 4 AU, a frozen planet with or without an atmosphere (if an atmosphere exists, it is expected to have a composition dominated by N_2 and CH_4). In the giant planets category, we expect, below 0.05 AU, a composition dominated by H_2, CO, N_2, and H_2O; between 0.05 and 0.1 AU, an atmosphere of H2, CO, NH_3, H_2O; beyond 0.1 AU, an atmosphere of H_2, CH_4, NH_3, H_2O.

What happens if the star is not of solar-type? Then the relationship between the distance to the star and the temperature is not the same as the Sun. Thus, for an A-type star, with a temperature of 10,000 K, the area of rocky planets with an atmosphere ranges from 1 to about 10 AU; in the case of an M-type dwarf, with a temperature of about 3200 K, the region is between about 0.05 and 0.5 AU. It should be borne in mind that the most important parameters in determining the nature of an exoplanet are its mass and its temperature. The albedo is also important, since it defines the amount of stellar flux absorbed and therefore the temperature of the object based on its distance from the star. Other factors also are involved, to a lesser extent: the rotation period, the existence or absence of a magnetic field.... Finally, we must remember that we have assumed a low-mass pro-toplanetary disk. What would happen in the event of a disk whose mass is one tenth of that of the star? Other formation mechanisms could possibly occur, as the gravitational contraction of a protosolar subcloud due to a local instability with the protosolar nebula. Perhaps one could also imagine rocky planets close to the star, massive enough to accrete the surrounding gas, who knows? The question is then whether, in the latter case, collision processes would allow the formation of such large rocky planets, or if they would be destructive, leading to the formation of multiple smaller rocky planets.

Of course, the above classification is no more than a very simple reading grid. It does not take migration into account, although we know this process to be common in exoplanetary systems. It is likely that the first atmospheric measurements of extrasolar planets will defy these predictions and lead to the discovery of new exotic objects!

6.5. From detection to characterization

Most of the exoplanets discovered to date have been found by indirect methods. The question is now to determine the physical nature of these objects, and to characterize their atmospheres. For this, a direct detection with spectroscopic analysis of the spectrum will be needed.

6.5.1 How to directly detect exoplanets?

The first method is to explore the direct imaging technique. It has already been successful, as we have seen, with the discovery of a very big exoplanet orbiting a brown dwarf (2MASS 1207, see Section 6.2.8.5)). The goal now is to get the image of less massive exoplanets closer to their star. We are back to the original question: how to overcome the stellar flux? A first method is the follow-up of the above observations, using adaptive optics, coronagraphy and increasingly sophisticated image processing techniques. This technique, first implemented by the astronomer Bernard Lyot, consists in obscuring the flux of the central object by a mask, in order to reveal patterns of low intensity at the periphery. Coronagraphy has been used, in particular, to study the solar corona during solar eclipses. In 1993, this technique led to the first detection of a dusty disk around the star Beta Pictoris. In 2008, coronagraphy allowed the detection of an exoplanet orbiting the star Fomalhaut. As of mid 2013, we know about thirty exoplanets detected by direct imaging. Several space projects (SEE-COAST, TPF-C, New World Observer) are under study, using the principle of coronagraphy in the visible and near infrared range.

An alternative method is nulling interferometry. The principle is as follows: the flux of the star is recorded by an interferometer in such a way that, at the central fringe, the flux is zero along the line of sight; this allows us to turn off the flux from the central star. If there is an exoplanet outside the line of sight but close to it, it can be detected in this way. Nulling interferometry is developed in the thermal infrared, around 10 microns. The space project using this principle is Darwin/TPF-I, in pre-study at ESA and NASA. The space mission would consist in a set of several telescopes working in interferometric mode. They would have to be located at a distance of at least 5 AU from the Sun, to overcome the zodiacal light of the solar system. This mission, very ambitious in terms of technology, has been postponed for some time due to feasibility problems, but could emerge again in a couple of decades or beyond.

A third method of direct detection consists in measuring the radio radiation of giant exoplanets, predicted to have a magnetic field as the giant planets of the solar system. A non-thermal emission is detectable in the field of decametric waves. The advantage of this method is that if we compare with Jupiter and the Sun, the non-thermal flux of the giant exoplanet is be comparable to the stellar flux. The radio radiation can be measured from the ground by large antennas or through the decametric array LOFAR, currently under development.

6.5.2 How to measure the spectra of exoplanets?

Direct observation of exoplanets and monitoring of their orbits may already provide valuable information on their seasonal temporal variation. If the measurements are accurate enough, it is in principle possible to determine the rotation period from their light curve. But in order to determine their atmospheric composition, spectroscopy, from the ultraviolet to the infrared range, is definitely necessary. How to achieve it? The most promising technique now seems to be the observation of the transit. We have seen that the measurement of transit (also known as primary transit), which corresponds to the passage of the exoplanet in front of its star,

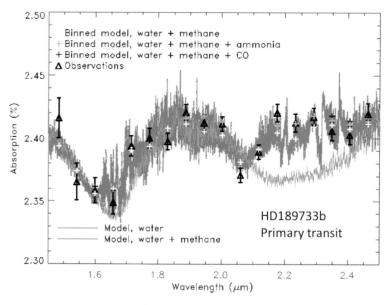

Figure 6.19. Primary transit of HD189733b recorded in the near-infrared range with HST-NICMOS, compared with two sets of atmospheric models, with H_2O and with $H_2O + CH_4$. It can be seen that the best fit is obtained when methane is present. After Swain *et al. Nature* **452**, 329, 2008.

Chapter 6. Exoplanets, the New Worlds

allows the detection of the object and the measurement of its radius. On this occasion, if the planet is surrounded by a large gaseous envelope, it is possible to detect some of its elements by absorption spectroscopy in front of the stellar flux. In this case, the atmosphere is observed at terminator (both morning and evening). Observations in the UV and visible range were recorded with the Hubble Space Telescope (detection of H I and C II in the UV, Na I, H_2 Rayleigh scattering, haze in the visible). Molecular species were detected in the neutral atmosphere in the near infrared range (H_2O, CH_4, CO_2). Another opportunity is provided by the observation of secondary transits, i.e. the passage of the exoplanet behind its star. By observing the stellar flux before, during and after the secondary transit, it is possible, by subtraction, to extract the spectrum of the exoplanet's day-side. Space data have been obtained with HST/NICMOS and with IRS and MIPS aboard Spitzer. Many articles have been published, leading to the (sometimes controversial) identifications of H_2O, CH_4, CO and CO_2.

So far, for feasibility reasons, observations of primary and secondary transits have been mostly limited to two very bright hot Jupiters, HD

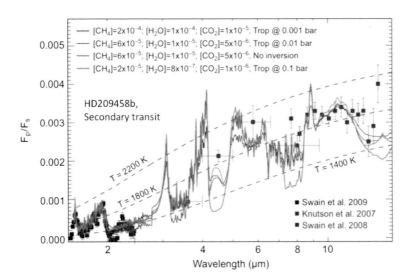

Figure 6.20. Emission photometry and spectroscopy data of HD209458b (secondary transit). Observations recorded by HST-NICMOS and Spitzer are compared with a set of atmospheric models corresponding to different CH_4 abundances and different thermal structures. After Swain *et al.* *Astrophys. J.* **704**, 1616, 2009.

209458b (see Section 6.2.8.1) and HD189733b. A few other sources have also been observed by spectroscopy, including XO-2b and GJ438b. There is already a lesson to be taken from these results: the observed atmospheric compositions of the hot Jupiters do not match the predictions of thermochemical equilibrium (see Section 6.4). Indeed, on both HD 209458b and HD189733b the temperature is high enough for carbon and nitrogen to be in the form of CO and N_2; still' CH_4 has been observed. There are two possible mechanisms which can induce a change in the atmospheric composition with regard to thermochemical equilibrium. The first one is dynamical quenching, which can bring species from the inside up to observable levels by vertical transport; the other is photochemistry, especially efficient in the case of methane. Both mechanisms are indeed present in the atmospheres of the solar system's giant planets.

Most likely, spectroscopic observations using both primary and secondary transits are going to develop in the coming decade, as

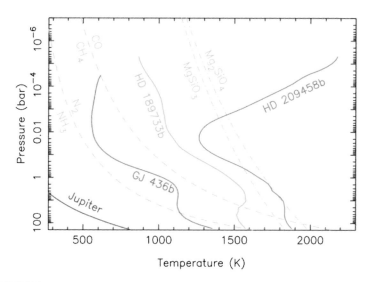

Figure 6.21. Thermal profiles of a few exoplanets, as inferred from transit measurements. The thermochemical equilibrium curves CO/CH_4, N_2/NH_3 and $MG_2SiO_4/MgSiO_3$ are also shown . They indicate in which form carbon, nitrogen and silicates are expected for a given range of temperature and pressure. The fact that CH_4 is observed on both HD209458b and HD189733b illustrates the effect of disequilibrium processes. The figure is taken from Tinetti *et al.*, EChO, Exoplanet Characterization Observatory, Proposal submitted to ESA, 2011.

Chapter 6. Exoplanets, the New Worlds

both techniques have their specific advantages and limitations. Primary spectroscopy is well suited for hot, inflated Jupiters transiting in front of small stars. Secondary spectroscopy is well adapted to hot Jupiters transiting around M-dwarfs. What is the most suitable spectral range for secondary transit studies? The thermal infrared range is the most favorable one, with respect to the contrast between the planetary and stellar fluxes. In the case of hot Jupiters whose temperatures are about 1200 K or higher, the heat flux extends toward shorter wavelengths down to the near infrared range, below 2 microns. Spectral signatures appear in emission or absorption, depending on the thermal profile of the exoplanet in the emitting region (troposphere or stratosphere).

The infrared range, between 1 and 30 microns, is the privileged domain for observing neutral molecules, which have easily detectable rotation or vibration–rotation bands. However, there are exceptions: molecular nitrogen has no infrared signature, but has strong emission bands in the ultraviolet. This is also the case for molecular hydrogen, also devoid of dipole moment, whose infrared signatures are weak. Atomic hydrogen has strong signatures in the UV (the most intense being Lyman alpha) and the visible (including the H alpha). The infrared spectrum of molecules depends on their structure. Molecules with a dipole moment have intense rotational lines to the far infrared and millimeter: this is particularly the case for H_2O, NH_3, HCN and CO. Those without a dipole moment (such as CO_2, CH_4 and most hydrocarbons) have signatures of vibration–rotation bands below 20 microns.

Now let us resume our simplistic classification of exoplanets and try to imagine what may be their spectrum, assuming again, as a first guess, thermochemical equilibrium. In the case of small rocky planets, we can find in the UV N_2, CO and CO_2 between 2 and 15 microns, and H_2O throughout the infrared spectral range. With the hope of detecting a planet where life has emerged, we also seek the signature of ozone at 9 microns, as we will discuss below (Chapter 7). In the case of small icy planets, we look again for N_2 in the UV, as well as methane, hydrocarbons and nitriles, between 1 and 15 microns. In the case of the giant planets, we look for H_2 and H in the UV, as well as CH_4 and hydrocarbons in the mid-infrared. If the object is not too cold, we will also look for the hydrogen-bearing molecules NH_3, PH_3 or H_2S between 2 and 11 microns, and H_2O throughout the infrared spectrum; in the coldest giant exoplanets, these molecules are likely to be trapped in the clouds below observable levels, as is the case for Uranus and Neptune.

The hardest part is yet to come: to measure the spectrum of exoplanets! So far, the measurements are still very patchy, and limited to very few sources. They were mostly made with the Spitzer and HST spacecraft, and have low spectral resolution, which makes the identification of molecules very difficult. Future progress will come from the use of new, more high-performance instruments on ground-based telescopes, and later from the extremely large telescopes (ELTs). The E-ELT, developed by the European Southern Observatory, is especially focused on the exploration of exoplanets, including super-Earths. Its first light should take place around 2022. Regarding space results, several projects are under study: the CHEOPS small mission has been approved by ESA for a launch in 2017, with the objective of defining precisely the masses and radii of transiting super-Earths, in order to define appropriate targets for future spectroscopy observations. PLATO, a space mission under assessment at ESA, is a follow-up to CoRoT and Spitzer, with special emphasis on bright and nearby stars. EChO, another project submitted to ESA, is devoted to the spectroscopy of transiting exoplanets. On the NASA side, the TESS mission, a follow-up of Kepler also designed for the study of bright stars, has been selected for a launch in 2016. Finally, the James Webb Space Observatory, a 6-m telescope designed as the successsor of the HST, is expected to operate by 2018. One of its objectives will be the spectroscopy of transiting exoplanets. In conclusion, the characterization of exoplanets' atmospheres has become a first priority of the astronomical community, and we can expect a lot of new results and exciting discoveries in the decades to come.

7

Searching for Habitable Worlds

Does life exist beyond Earth? This question has been an issue for mankind since the beginning of civilization; with the discovery of exoplanets, it has taken a new dimension. Today we are able to detect them, tomorrow we will characterize and study their atmospheres, their surfaces, and hopefully determine which ones could be habitable; ultimately, we will perhaps find possible signs of life. A new field of research is offered to the astronomy community which has started developing the tools and instruments necessary for this study.

7.1 A new discipline, astrobiology

The search for life in the Universe, whether in the solar system or on exoplanets, is not only a matter for astronomers, even though they are interested. The question of the emergence of life, and what conditions are favorable for it to occur, concern chemists and biologists; the problem of the emergence of life on Earth and the factors that made it possible concerns Earth-science physicists. Finally, the social dimension of the phenomenon is relevant to the humanities and social sciences. It can be seen that a wide community is implied in the search for extraterrestrial life. It merges into a new discipline, exobiology (also called astrobiology), whose purpose is the search for life and the study of all the conditions of its emergence. Multidisciplinary in nature, it includes astronomers, physicists, chemists, biologists and researchers in social sciences and the humanities.

The first question for the exobiologist community is: What is life? Biologists agree on the following definition. Living matter must meet three criteria: capacity for self-reproduction (i.e. identical reproduction), capacity for mutation (corresponding to an exception in the reproduction), and self-regulation against the environment (which allows its growth and conservation). The second criterion allows adaptation to changing environmental conditions (e.g. temperature or pressure changes). Species resulting from natural selection are those that are best suited to the new conditions.

Let's start with what we know: life on Earth. The cell is the structural basis of living systems. All living organisms use the same types of

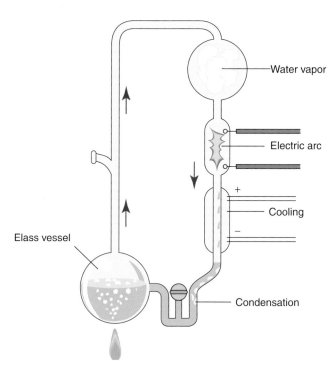

Figure 7.1. Diagram of the Miller and Urey experiment, which achieved the first laboratory synthesis of amino acids (after Ph. From Cotardière, Astronomy, 1999).

Chapter 7. Searching for Habitable Worlds

molecules, mainly nucleic acids and proteins. The main nucleic acid is deoxyribonucleic acid, better known as the DNA "double helix" which, while carrying genetic information, is responsible for the cell proliferation and cellular functions. Proteins are composed of twenty amino acids, called "prebiotic molecules" because they are the source of life without being themselves living matter. Remarkably, all these molecules have been detected in some meteorites, which proves they exist elsewhere than on Earth. The existence of an interstellar prebiotic chemistry has also been illustrated by a famous experiment by Stanley Miller and Harold Urey in 1953. It showed that it is possible to synthesize amino acids in the laboratory, in the presence of liquid water, from a mixture of reducing gases (H_2, CH_4, NH_3 ...) under the effect of electric shocks.

How did life appear on Earth? We have no answer to this question yet. We know that it first appeared in the oceans, more than 3.5 billion years ago. But how did it appear? Two assumptions are proposed: hydrothermal vents in deep oceans, where the synthesis of complex molecules does take place, and meteorite impacts, which could have enriched the atmosphere in amino acids. Anyway, three factors seem to have been very important for the development of life: the presence of liquid water, carbon and an energy source.

Could we imagine elsewhere in the Universe a form of life which would not use these basic ingredients? It cannot be excluded, but it seems unlikely. Carbon, with its four valences, is the most capable atom for creating links with other atoms. In the interstellar medium, where more than a hundred molecules have been discovered, most of them contain carbon, the most complex one being $HC_{11}N$. As shown in the periodic table of Mendeleev, the next atom with four valences is silicon, but it is ten times less abundant than carbon in the Universe, and we do not know any interstellar molecule containing more than one silicon atom. Thus carbon seems to be, by far, the best candidate species for generating a complex chemistry.

As for liquid water, it is also difficult to find a substitute. In fact, water is a molecule with a strong dipole moment; in the liquid phase, it is capable of dissolving polar organic molecules (known as hydrophilic molecules) while the non-polar molecules (known as hydrophobic molecules) do not react with water. The hydrophilic and hydrophobic properties of different molecules are used by living organisms for various special

Hydrogeneous compounds						
H_2	H_3^+					

Carbon chains and cycles						
CH	CH^+	C_2	CH_2	CCH	C_3	
CH_3	C_2H_2	$l\text{-}C_3H$	$c\text{-}C_3H$	CH_4	$C_4?$	
$c\text{-}C_3H_2$	$l\text{-}C_3H_2$	C_4H	C_5	C_2H_4	C_5H	
$l\text{-}H_2C_4$	HC_4H	CH_3CCH	C_6H	C_6H_2	HC_6H	
C_7H	CH_3C_4H	C_8H	C_6H_6			

H,O,C–bearing compounds					
OH	CO	CO^+	H_2O	HCO	
HCO^+	HOC^+	C_2O	CO_2	H_2O^+	
$HOCO^+$	H_2CO	C_3O	$HCOOH$	CH_2CO	
H_2COH^+	CH_3OH	CH_2CHO	HC_2CHO	C_3O	
CH_3CHOH	$c\text{-}C_2H_4O$	CH_3OCHO	CH_2OHCHO	CH_3COOH	
CH_2CHOH	$(CH_3)_2O$	CH_3CHCHO	CH_3CH_2CHO	CH_3CH_2OH	
$(CH_3)_2CO$	$HOCH_2CH_2OH$	$C_2H_5OCH_3$			

H,N,C–bearing compounds					
NH	CN	NH_2	HCN	HNC	N_2H^+
NH_3	$HCNH^+$	H_2CN	$HCCN$	C_3N	CH_2CN
CH_2NH	HC_3N	HC_2NC	NH_2CN	C_3NH	CH_3CN
CH_3NC	HC_3NH^+	C_5N	CH_3NH_2	C_2H_3CN	HC_5N
CH_3C_3N	C_2H_5CN	HC_7N	$CH_3C_5N?$	HC_9N	$HC_{11}N$

H,O,N,C–bearing compounds					
NO	HNO	N_2O	$HNCO$	NH_2CHO	$NH_2CH_2COOH?$

S,Si–bearing and other compounds						
SH	CS	SO	SO^+	NS	SiH	SiC
SiN	SiO	SiS	HCl	$NaCl$	$AlCl$	KCl
HF	AlF	CP	PN	H_2S	C_2S	SO_2
OCS	HCS^+	$c\text{-}SiC_2$	$SiCN$	$NaCN$	$MgCN$	$MgNC$
H_2CS	$HNCS$	C_3S	$c\text{-}SiC_3$	SiH_4	SiC_4	CH_3SH
C_5S	FeO	$AlNC$				

Deuterated species						
HD	H_2D^+	D_2H^+	HDO	CCD	DCN	DNC
DCO^+	N_2D^+	HDS	D_2S	NH_2D	ND_2H	ND_3
$HDCO$	D_2CO	$HDCS$	CH_2DOH	CD_2HOH	CD_3OH	CH_3OD
DC_3N	DC_5N	C_4D	CH_2DCCH	CH_2DCN	$c\text{-}C_3HD$	

Figure 7.2. Table of the gaseous interstellar molecules detected to date.

functions, such as producing a membrane. Are there other molecules capable of playing the role of water in the liquid state? One could think of ammonia NH_3, but the liquid phase occurs at lower temperatures, which would slow down chemical reactions. In addition, water has other advantages. As we have seen, this is a very abundant molecule in the Universe, which condenses at particularly high temperatures, and this is what made it a key element in the formation of planets. It is in liquid form in a very large temperature range. Finally, it has another remarkable property: water ice has a lower density than the liquid. This feature could have proved decisive for the history of life on Earth: in periods of cooling, the oceans were covered with a layer of ice, keeping the living species protected within the oceans. If the oceans began to freeze from the bottom, which is

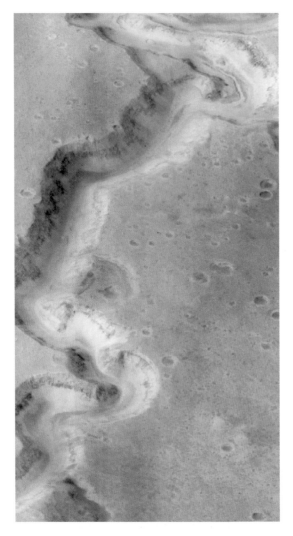

Figure 7.3. The Nanedi Valles region on Mars photographed by the Mars Global Surveyor. It is a good example of the Martian hydrology, sometimes very similar to that of the Earth (© NASA).

the case for many other molecules, living organisms present in the liquid would have been destroyed. It can be seen that liquid water has many advantages... and, in the lack of better candidates, we will use liquid water as a possible tracer for extraterrestrial life.

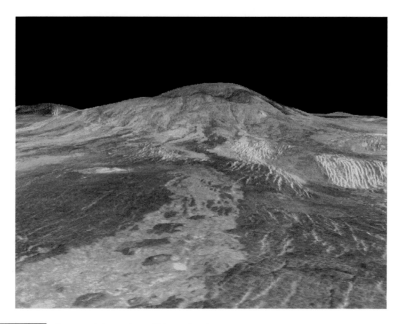

Images of the surface of Venus, synthesized from the Magellan radar mission. The surface of Venus is relatively Young (less than one billion years old) and covered with volcanoes (© NASA).

7.2 Life elsewhere in the solar system?

Before considering extraterrestrial life around other stars, philosophers and scientists of the XVIIth century have imagined it in the solar system, on Venus or Mars or even the Moon. This is especially the case of Fontenelle's claim in "The plurality of inhabited worlds." At that time, it was still pure speculation. At the end of the nineteenth century, the question came back on a scientific basis. The Italian astronomer Schiaparelli, in 1877, announced the discovery on Mars of linear structures which he called "channels". He was cautious about the interpretation of this result, but others were less so. In particular, the American astronomer Percival Lowell did not hesitate to see in these structures evidence of intelligent life: these channels were built by Martians to escape drought.

Other astronomers were skeptical: In the 1920s, Eugene Antoniadi, accumulating observations on his side, showed that the channels were the

Figure 7.5. The surface of Europa, observed by the camera of the Galileo spacecraft (© NASA).

result of an optical illusion due to the insufficient instrumental quality of previous observations. It took more time to kill the myth of intelligent life on Mars: the myth did survive until the arrival of the first spacecraft in the 1960s. But the debate is far from settled. Exobiologists are always looking for signs of life, past or present, on the red planet.

7.2.1 Life on Mars?

Why Mars? Because this planet is the most similar to Earth, with its sandy deserts, its volcanoes, its polar caps and seasonal cycles. Its carbon dioxide atmosphere is very thin and liquid water cannot remain on the surface under the current conditions of pressure and temperature. But that was not the case in the past (see Sections 4.4.2 and 4.4.3): we have a whole body of evidence suggesting that water flowed on Mars in the distant past, and that this atmosphere was denser and wetter. That is why, at the beginning of the space age, astronomers began looking for traces of life on Mars. The first mission devoted to this goal was the Viking mission, operated in the 1970s. It was composed of two identical orbiters and two surface

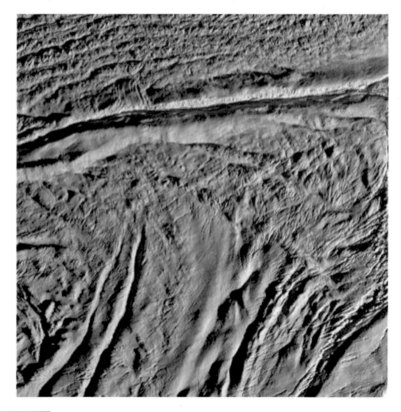

Figure 7.6. This region near the south pole of Enceladus, taken by the Cassini spacecraft in August 2008, shows faults and cracks indicating recent tectonic activity (© NASA).

modules, also identical. The overall mission worked perfectly, and remained in the annals of space exploration as a tremendous scientific and technological success; data accumulated over several years by all the instruments established a baseline still used today by the planetary community. Unfortunately, the surface module instruments detected no signs of life. Several experiments were designed to perform specific analyses of the soil outgassing in the presence of a nutrient medium, and to search for a possible photosynthetic activity of any carbonaceous constituents. In spite of some contradictory results, the instruments showed the total absence of organic molecules at the surface of Mars and the overall result was negative. The most plausible explanation is that the solar ultraviolet radiation, as it reaches the ground, destroys very quickly any organic molecule. So we should not look for traces of life on Mars in areas exposed to the Sun; this lesson from Viking was taken into account in the preparation of future missions.

After a decade marked by many failures, space exploration of Mars resumed in the late 1990s, alternating orbiters, descent modules and robotic vehicles. Multiple observations confirmed what the first surface images sent by Mariner 9 and Viking already suggested: large quantities of liquid water flowed at the surface in the past of the planet (see chapter 4.4.2). In particular, in 2002, the Mars Odyssey orbiter detected the presence of water ice under the perennial polar caps of Mars. Mars Express and the robot Opportunity revealed the presence of sulfates probably formed in the presence of liquid water; clays were found in the most ancient southern highlands. Water did flow on Mars, but when and for how long? Was the time sufficient to allow life to appear? This is the question that remains open. Future space missions will aim to search for sites where water could be liquid, hoping to find possible traces of fossil life.

7.2.2   A past ocean on Venus?

A priori, Venus is not a good candidate for exobiology. With its dense atmosphere of carbon dioxide, its hot temperature and sulfuric acid rains, it is one of the most inhospitable environments we can think of. However, early in its history, the planet must have been very different from what we know today. Let us remember that, at the beginning of its history, the radiation of the young Sun was lower than today: models of stellar evolution tell us that it was about 70% of its present value. Located at 0.7 AU from the Sun, the equilibrium temperature of Venus (assuming for the surface albedo a value of 0.3) had to be about 300 K, which would allow water (present in abundance at that time, see 3.2.2) to be in liquid form. But the inexorable rise of the solar flux gradually led to the evaporation of the oceans and water escape by photodissociation.

Unlike the Earth, Venus has retained no vestige of its ancient past, because volcanism has erased all traces of the surface before the last billion years. Therefore we will probably never know if Venus did ever harbor life.

7.2.3 An ocean under Europa's surface?

The atmospheres and surfaces of terrestrial planets are not the only places where one can find liquid water. There is another area highlighted by the phase diagram of water: the liquid phase is also possible at high pressure and temperatures (a maximum of several hundreds of bars and above 300 Kelvin, up to the critical point of water). We see these conditions in the interior of the outer solar system satellites, which are rich in water.

The first interesting case for exobiologists is Europa, the second Galilean satellite. As we have seen above (Section 5.3.1), the Voyager and Galileo spacecraft have provided several clues to the presence of a liquid ocean of salt water beneath the surface of ice. How to explain its presence? Tidal effects related to the presence of Jupiter, and its resonance with Io and Ganymede, provide enough energy, according to the models of internal structure, to maintain an internal temperature sufficient for water to be liquid. In the case of Io, the received energy is even stronger, but it had the effect of generating active volcanism that resulted in the loss of water. In the case of Ganymede and Callisto, models of internal structure also predict the possibility of a liquid ocean layer inside them, but, if it exists, it is probably trapped between two layers of ice. The interesting thing about Europa for exobiologists is that the ocean of liquid water, according to the models, could be deep enough to be in direct contact with the silicate core of the satellite. The implications could be significant for the development of a complex chemistry.

How to check the existence of the internal ocean and how to explore it? The problem lies in the thickness of the ice layer that covers it, which we do not know today. Depending on the model, it would be at least several kilometers and could reach several tens of kilometers. A recent result has brought a new hope: from the reanalysis of the Galileo data, a team of geologists have found that, in some places, underground lakes might be present just a couple of kilometers below the surface. In any case, the drilling of this ice is not for the near future. First, gravimetry and radar sounding experiments will attempt to confirm the existence of the ocean and to measure the thickness of the ice layer. This is one of the objectives of the JUICE (Jupiter Icy Moon Explorer) mission, recently selected by ESA. Other more ambitious steps will be considered for decades to come.

7.2.4 The icy satellites of Saturn

Other outer satellites could also have an ocean of water in their interiors. Enceladus, Saturn's satellite located near the E ring, appears to be in this category (see Section 5.3.3). The images and infrared measurements taken by the Cassini spacecraft have shown that the South Pole was subject to cryo-volcanic activity, resulting in the ejection of feathers. The spectroscopic analysis of these feathers has shown a large amount of water vapor. Located at four Saturnian radii from the planet, Enceladus is in 2:1 resonance with the satellite Dione. However, the energy dissipated by tidal effects is clearly insufficient, given the small size of the object. The active cryo-volcanism of Enceladus, which is probably the source of material that feeds the E ring, is still poorly understood.

7.2.5 A prebiotic chemistry on Titan?

Titan, as we have seen, is an exceptional outer satellite, with its dense atmosphere dominated by nitrogen and a surface pressure comparable to that of the Earth (see Section 5.3.2). Astronomers became very interested in Titan in 1981, after the discovery, by the Voyager mission, of a large number of hydrocarbons and nitriles. Other observations later completed the list with HC_3N and CH_3CN, two molecules whose presence is predicted by laboratory experiments in the synthesis of prebiotic molecules. For exobiologists, Titan appeared to be a potential prebiotic chemistry laboratory, perhaps comparable to what could have been the early Earth. There are, however, two important differences to consider. The first is the low temperature prevailing on Titan (93 K at the surface, 70 K at the tropopause), which significantly slows down all chemical reactions. The second is the atmospheric composition of the early Earth, which is not well known, but was probably based on carbon dioxide, far from the reducing atmosphere used in the experiments of amino acid synthesis. Nevertheless, the interest aroused by this unusual satellite was the driver for the ambitious Cassini–Huygens mission to explore the Saturn system (see Section 5.3.2). We expected to find a large number of new complex molecules. There has been almost no new detection of neutral species, but the surprise came from the ionosphere of Titan: the mass spectrometer of the Cassini orbiter has detected a large number of complex ions of hydrocarbons and nitriles, with atomic masses ranging up to several hundred. The most active complex chemistry does not take place in the neutral atmosphere, but in the ionosphere above it, at very high altitudes. These ions probably form complex condensates, feeding the yellow–orange aerosols that mask the surface of Titan. Their composition remains to be analyzed in more detail. The exploration of the Saturn system will continue with the monitoring of the Cassini orbiter and will probably bring new surprises. In the longer term, within a couple of decades, astronomers are considering a more ambitious exploration mission toward the Saturn system, TSSM (Titan and Saturn System Mission), which could involve a balloon to be sent into Titan's atmosphere and a close exploration of Enceladus.

7.3   Life on exoplanets?

We now leave the solar system, as well as *in situ* exploration. We will study exoplanets though remote sensing observations. We now have to define the best methods to identify the exoplanets that are most likely to harbor life, and to find diagnostics for highlighting the potential presence of life.

7.3.1 The habitable zone

Let us go back to the main criterion we have: the presence of liquid water. It depends on the temperature of the medium, and therefore several parameters that we defined above (Section 6.4): the distance to the star, the spectral type of the star and the albedo of the planet. Using an average value of the albedo (e.g. 0.3, a typical value in the solar system), we define for each star a "habitable zone": it is a ring around the star where the temperature is such that water can be present in the liquid state. For a solar-type star, the region typically extends from 0.9 to 1.3 AU. It

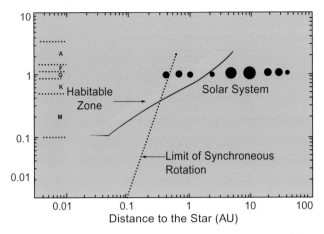

Stellar mass (in solar masses)

Figure 7.7. The habitable zone of a star according to its spectral type. The dotted line shows the limit within which planets are expected to be tidally locked (i.e. in synchroneous rotation with their host star) (© F. Casoli and Encrenaz T., ibid).

decreases down to 0.1 AU in the case of dwarf stars, and rises up to 10 AU in the case of the most massive stars.

We should mention immediately the limits of this notion of "habitable zone". Let us take the case of Venus, for example: due to its runaway greenhouse effect, the planet now has a surface temperature well above the equilibrium temperature corresponding to its distance from the Sun. We mentioned that liquid water had flowed on Mars in the past, and that Venus could also have hosted an ocean early in its history, yet the two planets are outside the habitable zone currently defined. There is another exception to the rule: we mentioned the "ecological niches" that could be found in some internal oceans of the outer satellites: they are far from the habitable zone of the solar system. The concept of habitable zone is thus nothing more than a first indication, certainly too simple, to guide us in our search for extraterrestrial life.

7.3.2 The search for biomarkers

Let us suppose that we have identified an exoplanet located in the habitable zone of its star. How to go further? The study of the chemical composition of its atmosphere is the key that can bring information on possible signs of life. To do this, exobiologists have defined what is called "biomarkers": they are simple molecules whose presence may be associated with the presence of life. They came to the conclusion that the simultaneous detection of CO_2, H_2O and O_3 would be a strong indication of the existence of biological activity. Why O_3? Because ozone comes from the photodissociation of O_2 and oxygen itself is not easily detectable spectroscopically. In contrast, ozone has a strong characteristic signature at 9.6 microns, easily detectable

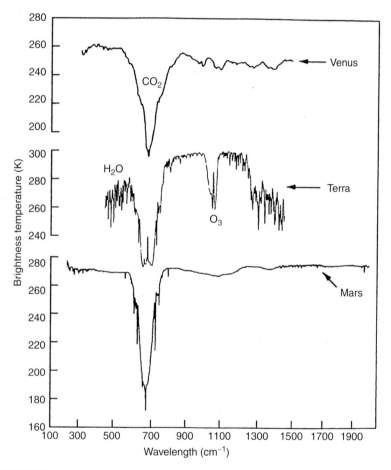

Figure 7.8. The thermal spectrum of the terrestrial planets. All three show the strong signature of CO_2 at 15 μm. On Earth, the spectral signature of the ozone is present at 1040 cm⁻¹, i.e. at a wavelength of 9.5 μm. Water is also present beyond 1300 cm⁻¹, at about 6 μm. After R. Hanel *et al.* Infrared Remote Sensing of the Solar System, CUP, 1992.

in the spectrum of the Earth; CO_2 and CH_4, are also spectroscopically very active and dominate the infrared planetary spectra (see Section 6.5).

Let us take another step: in order to find life on an exoplanet, the safest way would be to study its surface directly. Again, spectroscopy can help us. The spectrum of terrestrial plants in the near infrared, at about 0.7 microns, shows a characteristic signature, caused by chlorophyll, in the

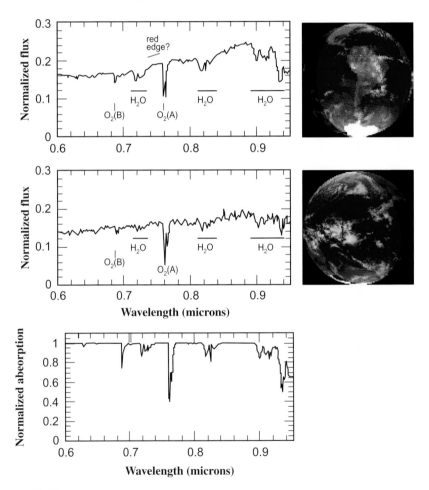

Figure 7.9. The spectral signature due to the presence of Chlorophyll at 0.7 μm was searched for in the earthshine from the Moon, i.e. the Earth's reflected light on the darks side of the moon. A slight difference between the spectra on the continent side (top) and on the ocean side (bottom) may be due to the presence of vegetation. According to Seager *et al.*, *Astrobiology* **5**, 372, 2005.

continuum spectrum: it is the so-called Red Vegetation Edge. Could we see such a signature on an exoplanet? The task is not easy, especially because of the likely presence of water vapor clouds that could partially mask the surface. As an exploratory observation, astronomers have attempted to measure the earthshine (i.e. the light emitted by the Earth reflected by the dark side of the Moon, observed at the time of the new moon), but the measurement is difficult. The detection of vegetation on the surface of an exoplanet requires, in addition to spectroscopy, imaging the disk, and not only measuring its integrated flux. Numerical simulations have synthesized the image of the Earth at a distance of 10 parsecs, with different spatial resolutions. Using a collecting area of 80,000 m² (the equivalent of a hundred telescopes of 30m each), a baseline of several thousand kilometers, and observing the visible light from the exoplanet for about half an hour, it would be theoretically possible to highlight continents and cloud structures from a distance of 10 parsecs. We can then obtain the visible spectrum of each pixel in order to find the vegetation signature. We are in the distant future, but not entirely in science fiction. Interferometry projects are in the pipeline, and astronomers hope they will emerge in the decades to come.

7.4 Searching for inhabited worlds

The search for possible extraterrestrial civilizations did not wait for the discovery of exoplanets. In the 1970s, astronomers Carl Sagan and Frank Drake tried to define by an equation the probability of existence of planets with technologically advanced civilizations in our Galaxy. N is the number of these planets. The equation can be written as follows:

$$N = N^* \times F_{pl} \times F_{habit} \times Fl_{ife} \times F_{civil} \times <T_{civil}>/T^*$$

In this equation, called the Drake Equation, N^* is the number of stars in our galaxy, F_{pl} is the fraction of them hosting a planet, F_{habit} is the fraction of them in which life appeared; F_{civil} is the fraction of them having reached a level of civilization capable of communicating; $<T_{civil}>$ is the average lifetime of a technologically advanced civilization, and T^* is the lifetime of stars.

Chapter 7. Searching for Habitable Worlds

We know N * (about 10^9) and T * (about 10 billion years for a solar type star). Since the 2000s, we have had an estimate of F_{pl}. However we do not know any other factor!

Observed from the outside, the radio signals emitted by the Earth show the presence of a technologically advanced civilization. For several decades, radio astronomers have studied this possibility to try to communicate with potential extraterrestrial civilizations using radio signals. To carry their messages, they selected the transition of atomic hydrogen at a wavelength of 21 cm, observable through the Earth's atmosphere, and they have begun to continuously monitor the sky at this wavelength, using the most powerful radio telescopes. The SETI (Search for Extraterrestrial Intellligence) project was born in the United States in 1984 on the basis of this concept. Initially funded by NASA, it now operates with private funds through an extensive international collaboration. Another research project called CETI (Communication with Extraterrestrial Intelligence) aims, not to receive messages from outside, but to emit signals toward external civilizations. Frank Drake, using the Arecibo radio telescope in 1974 sent a coded message towards the globular cluster M13.

All these studies have so far been unsuccessful, but they reflect the collective interest in the search for extraterrestrial life, an interest that keeps growing and goes beyond the scientific community. Another example is given by the messages carried by the Voyager probes, which now travel at more than 100 AU from the Sun, and continue to move away from the outer solar system.